Advance Praise for *Outsmart Your Brain*

"In a sentence, this is the best book I've read on how anyone can learn the tactics of the most successful students. Practical but backed by the latest science, *Outsmart Your Brain* is an on-ramp to the virtuous cycle of interest, confidence, and achievement."

—Angela Duckworth, *New York Times* bestselling author of *Grit*

"If left to our own devices, humans will usually study poorly. Luckily, Dan Willingham has identified all the ways we can trick our brain into learning (before it tricks us first). A user's guide to the student's brain."

—Amanda Ripley, *New York Times* bestselling author
of *The Smartest Kids in the World*

"The ultimate guidebook for doing well in school, and the perfect gift for any student heading off to college, and also for the high school student who is beginning to care about grades, or—better yet—actual learning."

—Jonathan Haidt, *New York Times* bestselling coauthor
of *The Coddling of the American Mind*

"Brisk and interesting, this is a wonderful book with a wealth of practical advice for students in 'how to' chapters on many topics. I would also recommend the book for teachers and lifelong learners—anyone who cares about learning."

—Henry L. Roediger III, coauthor
of *Make It Stick: The Science of Successful Learning*

"Willingham does double duty: he places the power to learn back where it should be, in the hands of students, while showing teachers how to harness the most effective systems and techniques for boosting learning. An essential tool for the new school year, every bit as important as that calculator, handful of sharpened pencils, and stack of notebooks."

—Jessica Lahey, *New York Times* bestselling author
of *The Gift of Failure*

Also by Daniel T. Willingham

Why Don't Students Like School?
The Reading Mind
Raising Kids Who Read
When Can You Trust the Experts?

Outsmart Your Brain

WHY LEARNING IS HARD

AND HOW YOU CAN MAKE IT EASY

Daniel T. Willingham, PhD

G

GALLERY BOOKS

New York London Toronto Sydney New Delhi

G

Gallery Books
An Imprint of Simon & Schuster, Inc.
1230 Avenue of the Americas
New York, NY 10020

First Gallery Books hardcover edition January 2023

GALLERY BOOKS and colophon are registered trademarks of Simon & Schuster, Inc.

For information about special discounts for bulk purchases, please contact Simon & Schuster Special Sales at 1-866-506-1949 or business@simonandschuster.com.

The Simon & Schuster Speakers Bureau can bring authors to your live event. For more information or to book an event, contact the Simon & Schuster Speakers Bureau at 1-866-248-3049 or visit our website at www.simonspeakers.com.

Interior design by Davina Mock-Maniscalco

Manufactured in the United States of America

10 9 8 7 6 5 4 3

Library of Congress Cataloging-in-Publication Data

Names: Willingham, Daniel T., author.
Title: Outsmart your brain : why learning is hard and how you can make it easy / Daniel T. Willingham.
Description: First Gallery Books hardcover edition. | New York, NY : Gallery Books, [2023] | Includes bibliographical references and index.
Identifiers: LCCN 2021045800 (print) | LCCN 2021045801 (ebook) | ISBN 9781982167172 (hardcover) | ISBN 9781982167219 (ebook)
Subjects: LCSH: Learning, Psychology of. | Study skills—Study and teaching. | Cognitive learning. | Motivation in education.
Classification: LCC LB1060 .W5434 2022 (print) | LCC LB1060 (ebook) | DDC 370.15/23—dc23/eng/20220215
LC record available at https://lccn.loc.gov/2021045800
LC ebook record available at https://lccn.loc.gov/2021045801

ISBN 978-1-9821-6717-2
ISBN 978-1-9821-6721-9 (ebook)

This book is dedicated to
Sherry Willingham Segundo
and Judy Willingham Shimm

CONTENTS

Introduction   1

1. How to Understand a Lecture   7
2. How to Take Lecture Notes   27
3. How to Learn from Labs, Activities, and Demonstrations   50
4. How to Reorganize Your Notes   74
5. How to Read Difficult Books   90
6. How to Study for Exams   105
7. How to Judge Whether You're Ready for an Exam   135
8. How to Take Tests   149
9. How to Learn from Past Exams   174
10. How to Plan Your Work   193
11. How to Defeat Procrastination   216
12. How to Stay Focused   237
13. How to Gain Self-Confidence as a Learner   259
14. How to Cope with Anxiety   273

Conclusion   289
Acknowledgments   295
Bibliography   297
Index   315

When you started preschool, your teachers and parents had no expectation that you would be responsible for your own learning. No parent has ever said to a five-year-old, "Your teacher tells me that you're not really giving your best when it comes to learning your colors. She also says that you don't fingerpaint like you really *mean* it. I don't see why I should keep paying for preschool if you're not going to apply yourself!" It was your teacher's responsibility to create an environment where you would learn.

But by your early teens, school had morphed into a format where you carried much greater responsibility for your own learning. The teacher lectured while you took notes; at home, you read textbooks, completed assignments, and studied for tests. This class format meant your teachers expected that you knew how to (1) set priorities and plan your schedule; (2) read difficult content independently; (3) avoid procrastination; (4) memorize content; (5) avoid distractions; (6) judge when you had studied enough; (7) show what you knew on a test; and (8) deal with emotions like anxiety that interfere with learning. And if you didn't do those things well, it was your problem, not the teacher's. In short, you were expected to be an independent learner.

But your brain doesn't come with a user's manual. Independent learning calls for many separate skills, and you needed someone to teach them to you. Most likely, no one did. Surveys of college students show that the vast majority devise their own strategies for studying, avoiding procrastination, and so on. But the strategies they come up with usually aren't very good. That's why I wrote this book. It's a user's guide to your brain that will allow you to fully exploit its learning potential and so become an independent learner.

How I Came to Write This Book

My primary motivation in going to graduate school was not an altruistic wish to help people learn but a selfish wish to become a professor, because I believed that professors didn't have bosses. (That turned out to be less true than I thought but more true than I probably deserve.) I entered a psychology doctoral program with a let's-see-how-this-goes attitude, which is exceptionally stupid planning.

But I got lucky. I found myself fascinated by the human mind and especially human learning. I finished the program with enthusiasm, and I lucked into a job teaching at a college. My research concerned memory, but it was pretty technical and removed from everyday life. You've heard the joke about the guy who gets a PhD, whereupon his mother explains to her friends, "He's a doctor, but not the type who helps people." I was a learning researcher, but not the type who helps you learn.

So it went for about ten years. One day I got a phone call from a near stranger, inviting me to come to Nashville to deliver a lecture on learning to five hundred teachers. I politely pointed out that I didn't know anything about teaching because I was a doesn't-help-you type of researcher. He said, "Sure, we get that. We just think teachers would find it interesting." Puzzled but flattered, I said, "Okay."

Six months later it was time to write the talk, and I panicked. Obviously teachers know how children learn; what could I possibly say that they didn't already know? I considered backing out, but I knew it was too late for the event organizers to replace me. I threw together a fifty-minute talk, plucking a few ideas from the introductory course on cognition that I had been teaching to college sophomores. I was so certain the talk would flop that half an hour before it started I asked my wife (a teacher), whom I had dragged to Nashville for my first talk about teaching, *not* to attend.

But to my considerable surprise, it was a success. Teachers didn't know the content, even though it covered material you'd take in your very first course on learning. Furthermore, they saw it not as abstract but as useful in their classrooms.

My career changed course. I thought teachers could benefit from knowing what scientists have figured out about how people think and learn, so I started writing articles and books that explained it.

I also started thinking about how this information applied to my own students. I added a "how to study" lecture to my introductory course on cognition. Students said it was useful, but their grades didn't change much. I had focused on efficient ways to commit information to memory, so I guessed that there must be other aspects of studying that caused problems.

When students came to my office for help, I started asking more questions about their study habits and strategies. I asked them to bring their textbooks and notebooks to our meetings so we could talk about how they read and took notes.

I learned that my students struggled for many reasons, not just poor memorization strategies. Some didn't know how to comprehend a complex book chapter, some procrastinated, some had trouble understanding lectures, some choked when they took a test, and so on.

After about a year I felt I was getting pretty good at diagnosing where the problem lay for any given student. But I wasn't great at getting students to change how they studied, which, to be honest, I thought was strange. They came to me because they knew things weren't going well. Why not try my advice?

Why Your Brain Must Be Outsmarted

I solved the puzzle by accident. A student asked me how I had become interested in memory, and I was reminiscing about a course I had taken in graduate school. "I was so struck by the *weirdness* of memory," I said. "So much of what I thought was true wasn't." As the words were coming out of my mouth, I realized how strange my advice about studying probably sounded to my students.

For example, *wanting* to learn has no direct impact on learning. You often remember things you didn't try to learn. I expect you could tell me whether or not Prince Harry is married, what Harvey Weinstein did wrong, and whether or not Bradley Cooper played the lead in the movie *Forrest Gump*. You didn't study any of these things; you were simply exposed to them, and they stuck in your mind. When I was a college student, I spent much time frantically trying to cram new knowledge into my head; it was weird to be told that the desire to learn doesn't matter.

I was equally dumbfounded to discover that repetition, although it often helps learning, doesn't guarantee it. For example, do you know what's written across the top of a dollar bill? There's an eagle on the back of the bill; what appears over its head? Given the number of dollar bills you've seen in your life, with all that repetition, you'd think you'd know what one looks like.

So I started asking my students, "Please, be honest: Did you try any of those strategies I recommended?" Most said they had, but not more than once. The problem wasn't that the strategies sounded weird, it was that they felt ineffective while they were doing them.

That made sense to me; learning is like exercise in this way. If you want to increase the number of push-ups you can do, you could practice push-ups, but it would be even better to practice really difficult push-ups, like those where you launch yourself off the floor and clap. You can't do very many of them, so it feels counterproductive. "This is stupid. I'm trying to do a lot of push-ups, and I can only do a few of these!" You have to keep in mind that the greater challenge will make you stronger in the long run. In contrast, if you practice push-ups on your knees, it *feels* like things are going great because you can do so many so quickly, but it's obviously a less effective exercise.

When you're trying to learn, your brain tells you to do the mental equivalent of push-ups on your knees. Your brain encourages you to do things that feel easy and feel like they are leading to success. That was why my students, left to their own devices, drifted toward the same ineffective study strategies. Outsmarting your brain means doing the mental exercise that *feels* harder but is going to bring the most benefit in the long run.

How to Use This Book

Most of schooling—starting around age twelve and continuing through postcollege education, like medical or law school—has the same format: You learn by attending lectures and reading on your own. You demonstrate your learning by taking tests. There's more to schooling than that (sometimes you have to write a paper, for example), but these three tasks—listening, reading, taking tests—make up the bulk of a student's work. So these are the tasks I've addressed in the book.

Naturally, each of these basic tasks has subcomponents. For example, studying for a test requires not only committing things to memory but having good notes to study, planning time in your schedule to study, and so on.

Each chapter of this book guides you to success in one of these processes. You can pick and choose which chapters to read according to which aspects of learning you want to improve. You don't have to read the chapters in order or read all of them. And I don't expect that you will use all of the tips in a chapter. I offer a bunch so you can select one that appeals to you; if it doesn't work, try another. But don't reject a strategy simply because it sounds to you as though it won't work. Remember, many will sound funny, and they may feel, at the time, as though they're not working! Judge the effectiveness of a method by the results, not by how it feels to do it. Instructors will find the advice for students useful, but there's also a section at the end of each chapter that describes how they can make use of the same principles in the classroom.

Your memory is a tool, and this book is an operating manual that will allow you to become an independent learner. I can't promise that I'll make learning completely effort-free. The brain just doesn't work that way, and if anyone tells you otherwise . . . well, keep your hand on your wallet while they're around.

What I can promise is much greater efficiency. I will show you how to change your approach to learning so that you can learn on your own and so that the effort you put in will have much greater impact. You'll learn faster, and what you learn will stick with you longer. All you need to do is understand a bit about how your brain works—and about its stumbling blocks. Then you can outsmart it.

CHAPTER 1

How to Understand a Lecture

By the time students get to college, they've listened to thousands of hours of lectures, so you'd think that they'd all be quite good at learning that way. They usually aren't. Part of their problem is the inability to take good notes, and I'll tackle that topic in the next chapter. Here I want to focus on understanding what the instructor says.

Now, if you don't understand, your next step seems obvious: ask for clarification. But what if you fail to understand and *you don't realize that you haven't understood*? How are you supposed to guard against that?

Let's consider the process of noticing that you don't understand something. That feeling is triggered by a failed search of your memory. For example, a talkative stranger at the grocery store says, "Wow, this stack of cans is in a parlous state, right?" Or a friend asks, "What does it mean when a bird sings at night?" In either case, you search your memory for information (definition of *parlous*, why insomniac birds sing), you don't find it, and so you think, "I don't get it."

There's a second type of failed memory search that leads to confusion, and it's based on how people communicate. When people speak, they don't say a lot of what they actually mean. They are not trying to be mysterious; they assume that you have the missing information in your memory and will use that information to fill the gaps in what they said. For example, suppose a friend says:

> "What the heck, I called Domino's an hour ago. Have you seen my phone?"

The connection between the first and second sentences seems obvious—the friend is asking about his phone to call the pizza place—but consider how much information is needed to make that connection. Your friend assumed you knew that Domino's is a business that delivers pizza, that you knew an hour is a long time for pizza delivery, that calling the store is an appropriate action for poor service, and that phones are for making calls.

We always omit information when we speak. If we didn't, communication would be really long and really boring. ("Toss me my phone, would you? Because I want to make a phone call, and that's what phones are for.")

Now imagine your friend says this:

> "What the heck, I called Domino's an hour ago. There are at least six minnows in the shallow part of the pool."

It's fine for neighboring sentences not to have an obvious connection—sometimes someone's talking about pizza, and the next moment she's asking about her phone—but we assume we will find a connection once our memory is consulted.

So we recognize that we've failed to understand when we probe our memory for either (1) a fact (the meaning of *parlous*) or (2) a connection (pizza and minnows) and find nothing. These are cases when you know that you don't understand and you can do something about it—most obviously, ask the speaker to explain.

Now, when would you fail to understand and not even know that you're missing something?

It won't happen with an unknown vocabulary word, but it could with a connection, because there can be more than one possible connection. Perhaps you connect two ideas in one way and hence think you've understood. But the speaker thought you would *also* connect them in another way. You've missed something, but you don't realize it.

For example, suppose in a history class the instructor says:

"A lot of movies starring Shirley Temple came out during the 1930s. They were meant to make their audience feel good and forget their troubles."

A listener might think that he's understood the connection between the sentences: each provides a fact about Shirley Temple movies. But suppose that a few days earlier the instructor had taught about the Great Depression: that economic times were terrible in the 1930s and most people were struggling financially. The lecturer thought listeners would understand that Shirley Temple's movies were popular because they made people feel good during economically difficult times.

So now we see how you might fail to understand something but not perceive that you don't understand: you make a connection between ideas, so you think you've got it, but the instructor wants you to connect them in another way.

This sort of problem is especially likely to pop up during lectures because of the way they are organized. Conversations are unplanned; I just talk about things as they occur to me, so connected ideas typically follow one after the other almost immediately. But lectures are usually organized hierarchically, which means the instructor wants the listener to connect some ideas that are not next to one another. Let's look at what that means.

Imagine taking a food science class and attending a lecture on cooking meat. There are three main topics for the day: cooking meat kills bacteria, it imparts flavor, and it makes meat more tender. The figure on the opposite page shows a partial outline of the lecture.

This is the organization the speaker might have in her head, but it's not the organization you would experience in her class. No one talks in a hierarchy. Learners experience lectures linearly. The capital letters show the order in which a speaker would talk about each point.

The ideas labeled A, E, and L ("kills bacteria," "flavor," and "tenderness") ought to be linked. Those all exist in a subcategory: the three reasons that humans cook meat. But if the instructor simply goes through the lecture without highlighting that, some listeners will miss that important connection. The neighboring sentences in the lecture will probably connect well enough, so that there's no sentence that surprises students and makes them wonder, "Wait, what is this idea supposed to connect to?"

Now we see why most students get the factoids in lectures, for example, the definition of terms such as *collagen* and *psoas muscle*. They notice they don't know those words, just as you noticed *parlous*. It's the deeper connections they miss, ideas that are related by how they function or because they are all evidence for or examples of a broad conclusion. The information they miss is also the information instructors think is more important.

In summary, your brain evolved to understand typical speech. In a normal conversation you don't plan fifty minutes of remarks in advance; you say things as they occur to you, and because you're planning only a sentence or two at a time, you're unlikely to say something that can be understood only if your listener connects what you're saying now to what you said twenty minutes ago. But lectures are planned and organized hierarchically. Therefore, it's not just possible that an idea connects to something mentioned twenty minutes ago, it's likely, and if a student misses that connection, she will miss a layer of meaning.

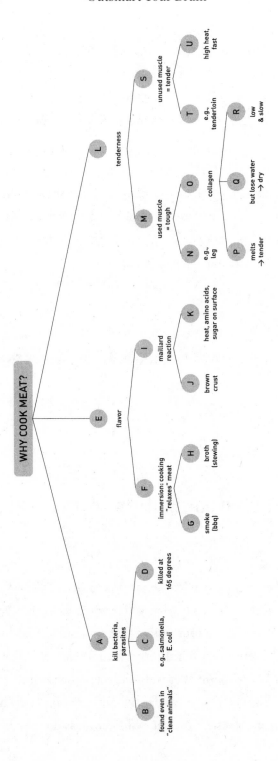

WHEN LEARNING BY LISTENING
What your brain will do: It will listen to a lecture the way you listen to a friend speaking and therefore miss deeper connections in the content.
How to outsmart your brain: Plan for the mismatch between the way the speaker thinks of the content being organized (a hierarchy) and the way you experience a lecture (linearly), so that you make the connections the speaker wants you to.

In this chapter you'll learn some tricks to ensure that you get the deeper meaning of a lecture, not just the new vocabulary words and factoids.

TIP 1

Extract the Organization from a Lecture

Ideally a speaker will be explicit about organization; she will tell you at the start of the lecture, "This is what you're going to learn. The main conclusion is X. There will be four points that support X." And then during the lecture, she'll refer back to this organization, saying, "Okay, now we're finished with the first point that supports our conclusion. Let's move on to the second." She's telling you what the organization is as she goes.

But what if she doesn't? In that case, you must do your best to figure it out yourself. For example, in the lecture we discussed earlier about cooking meat: if the instructor says, "Cooking also makes tough meat more tender," you're supposed to know that this statement is one of the three reasons explaining why humans cook meat.

But as you listen, you're not going to appreciate every bit of the lecture organization. It moves too quickly. **Aim for getting the first two levels of the hierarchy.** The top level is the question, or the overriding theme of the day. In our food science lecture, the top-level question was "Why cook meat?" In a history class, it might be whether presidential candidates today could run front-porch campaigns—candidacies in which the politician does not travel but makes speeches close to home.

You may get help in determining the lecture organization from a written document you see before the lecture—a syllabus if it's a class, a handout if it's a presentation—that gives you some idea of the main topic. If you have no advance knowledge of the theme, a decent guide is *whatever the speaker says first.* Speakers almost always provide a summary, even if it's just a sentence or two, of the topic to come. Which means that if you're a minute late, you'll miss it. If you're slow to turn your attention to the speaker because you're chatting with the person next to you or you're on your phone, you'll miss it. **Be present and ready for the start-of-presentation summary.**

The second level of the hierarchy will be pieces of evidence that support the conclusion of the day. As we saw in the food science lecture, it was the three reasons people cook meat. In the history class, perhaps the second level of the lecture would be examples of successful (and unsuccessful) front-porch campaigns, the nature of media when such campaigns were conducted, the characteristics of the candidates who ran them, and then a summary of these factors as related to modern politics.

If the point of the lecture is to teach you to *do* something—draw blood, for example—the subpoints might be substeps of the procedure, justifications for why they are effective, or a list of circumstances describing when to use each method.

Again, a good instructor will use verbal cues that explicitly say, "I've finished defining the features of a front-porch campaign, so now I'll give some historical examples." Ineffective instructors won't do that, but *they* know they are shifting topics, even if they don't think to tell you. So **listen for verbal cues that provide clues to organization**, for example:

- "The second reason . . ."
- "That raises a different question."
- "So *now* we know . . ."
- "Let's look at this from a different perspective."
- "Anyway . . ."
- "Okay."

Look for nonverbal cues. Instructors usually stop for questions when they've finished covering a topic, to be sure their listeners understand it before moving on to something new. If the instructor stops to consult her notes, or even if she pauses for a moment to think, that probably signals a shift to a new topic; she's finished with one idea and is checking to see what's next.

You shouldn't try to put together the whole hierarchy while you're listening to a lecture, but do **try to interpret details in light of broader ideas**. Remember, the whole point of this chapter is how to understand new content as you listen. Part of understanding is interpreting things in the right context. For example, take the fact that when James Monroe was elected president in 1820, he received every vote in the electoral college except one. This same fact might be mentioned:

- As evidence that it was an "era of good feeling" and harmony in the United States

- As evidence of the weakness of the Federalist Party after the War of 1812
- In the context of Monroe's hope that the party system would die out

To interpret details in light of the big picture, you have to keep the big picture continually in mind as you're listening. That's hard to do because you're trying to follow the lecture and take notes. So instead, mentally check in with the big picture every now and then. Suppose you've learned about vectors in a previous class and now the instructor introduces the idea of vector addition. It's hard to simultaneously understand this new idea and think about how it connects to other ideas in the course. So try to think about it when the instructor is ready to shift to a new topic. When the instructor asks if there are any questions, don't just ask yourself, "Do I understand what she just said?" Also ask yourself, **"Do I understand how what she just said relates to the broader topic of the day?"** If it's not obvious, ask.

> *In a sentence:* Expect that lectures will be hierarchically organized, and try to extract the organization during the lecture.

TIP 2

Expect Listening to Require Work

People often mistakenly think that attending a lecture is easy because they're just listening. In fact, lectures have a bad reputation among some educators because they seem passive; students just sit there. But this is

inaccurate, and in the previous section we saw an important reason that
learning from a lecture requires active thinking: listeners must rebuild
the hierarchical organization of what they hear.

There are other important ways in which lectures differ from typical
conversation. People use more unusual vocabulary when they deliver a
lecture, and they are communicating more difficult ideas than you nor-
mally would when talking to a friend. Further, your friend usually notices
whether or not he is understood; he might pause or say, "You know?"—
which is your cue to show that you're getting it by nodding or saying,
"Right." Instructors pause for questions much less frequently.

Plutarch, the Greek biographer, commented on the difficulty of listening nearly two thousand years ago:

> There are others who think that the speaker has a function to
> perform, and the hearer none. They think it only right that the
> speaker shall come with his discourse carefully thought out and
> prepared, while they, without consideration or thought of their
> obligations, rush in and take their seats exactly as though they
> had come to dinner, to have a good time while others toil. And
> yet even a well-bred guest at dinner has a function to perform,
> much more a hearer; for he is a participant in the discourse and a
> fellow-worker with the speaker.

I've taught a large lecture course for each of the last thirty years, and
I've been speaking to groups of adults at schools and corporations for the
last fifteen. Unengaged students and adults look the same, and they are
easy to spot. They slump in their seats. Their eyes are dull, and they focus
only slowly when I start to talk. It's not that they are tired or anxious or
distracted by personal problems; it's that they are *passive.* They're treating
a lecture like a movie or a concert.

It's easy to see why you'd feel like you're part of an audience when you're in a large lecture hall with a few hundred other students. It's natural to expect that the entertainment will come to you. But you'll fare much better if you come to each lecture psychologically prepared to put in some mental effort.

> *In a sentence:* Learning by listening takes work, so come to each class with that expectation. *Don't be at all passive about it.*

TIP 3

If You're Given Notes, Use Them to Check Your Notes, Not Replace Them

Suppose the speaker provides you with copies of her notes. Or with an outline of the lecture or copies of the figures. How should you use them? You can get closer to answering that question by answering another: Why do you take notes in the first place?

Researchers have asked people that question, and they point to two functions that you've probably thought of: First, just writing things down makes them more memorable. Second, reading your notes later jogs your memory. Research shows that notes do serve both functions.

Now consider how each function is affected by getting notes from the instructor. We might guess that those notes will be more complete and accurate than the notes you take. In fact, they will probably have all the deep connections that I explained are hard to capture while you are listening. So they would seem to be quite good for the memory-jogging function. But you won't get the memory boost that comes from writing things down. The instructor did the writing, not you.

Our guess—that using instructor notes is both a plus and a minus—matches what researchers have found. There's not a clear advantage to learners taking notes versus being given notes. That may be why some instructors don't provide notes—they don't see the point.

But suppose you do get notes or an outline or slides. What should you do with them? Although there's not a clear, research-based answer, we can make a reasonable guess, based on the two purposes of notes.

You still want the memory benefits that come from taking your own notes. So **take your own notes, even if you know you will get notes later**. And if you get them before the lecture, don't bring them with you, figuring you will follow along and add your own observations to them. You're not going to get the same memory boost, and trying to simultaneously listen to the lecture and line it up with the written outline can get confusing. The same goes for PowerPoint slides: don't print them and take notes on them.

If you get notes or an outline before the lecture, look them over. You don't need to spend a long time doing it. Just **identify the top two levels of the hierarchical organization of the lecture**: What's the overall theme, and what are the main subpoints?

Knowing this information in advance provides a big advantage to your comprehension and your note taking. Write the theme and subpoints at the start of your lecture notes for easy reference. Then, as the lecture progresses, you'll know where you are in the overall lecture organization, and you can mark it as you go.

You'll still want to **coordinate your notes with the instructor's notes later**. That's obviously your only option if the notes are available only after the lecture, but even if you get them beforehand, afterward is the time they'll prove most useful. The process of working with your notes after you take them is so important that I devote all of chapter 4 to it.

> *In a sentence:* If the speaker provides notes or an outline, use them to aid your comprehension before or after the lecture, but don't consider them a replacement for your own notes.

TIP 4

Be Thoughtful About When to Read Assignments

There's often some assigned reading associated with a lecture, and you're supposed to show up having read it. The logic "read first, then listen" seems obvious; you will understand the lecture better if you already know something about the topic. Recall that when people write or speak, they exclude some information that their audience needs for understanding, on the assumption that listeners have that information in memory. That was the point of the example concerning Shirley Temple movies; the instructor assumed that the students knew that the Great Depression had occurred during the 1930s and that they would conclude that the economic circumstances primed people to enjoy this kind of film. You will understand more if you already know something about the topic, so doing the reading first will help you understand the lecture.

But it turns out that the reverse is equally true. If you go to the lecture and then do the reading, you'll understand the reading better.

Making the right decision—reading first or lecture first—really depends on what the instructor assumes you know when you walk into the lecture hall. On the one hand, if you diligently do the reading first and then the speaker explains all the content you read but is clearer than the book was, there was obviously no reason to do the reading in advance. On the other hand, if you don't do the reading and the lecturer assumes you know that content and goes beyond it, you will definitely be confused.

The key to answering the question "Should I do the reading before or after the lecture?" is knowing what the instructor assumes you've gotten out of the reading before you come to the lecture. Of course, you can simply ask the instructor what they expect. They will likely say they want you to have done the reading beforehand. Still, they may not *teach* that way.

For example, when I was in college, I took a course in epic poetry: we read the *Iliad*, the *Odyssey*, *The Song of Roland*, and several other works. I found each of them pretty difficult to understand, and I don't mean anything very deep by *understand*; I mean I had trouble following what happened in the poem. We were to come to class having read something like fifty pages, and the professor would lecture, focusing on historical and cultural information that helped put that bit of the poem into context.

About the third week I noticed that the instructor started each session with a summary of the reading; he'd provide the basic outline of events in three minutes. So I started doing the reading after I attended class. Having his skeletal summary in mind made it much easier for me to comprehend the poem. And not having read it before class didn't affect me much because he had provided the summary, so I could more or less follow the historical and cultural material.

If you find the instructor's lectures quite easy to follow but find the readings difficult, try reading after the lecture and see if it helps.

In a sentence: The second time you encounter material, it's easier to understand, whether you're reading it or hearing it; plan your reading and listening accordingly.

TIP 5

Get Over Your Reluctance to Ask Questions

Earlier in the chapter, I described how a failure to understand can slip by you. But other times you know darn well that you don't understand. If it happens while you're listening to a lecture, the solution would seem to be simple: stick your hand in the air and say, "Wat?" For many people, it *is* that simple. But others are reluctant to ask questions, usually because they (1) "don't want to be annoying," (2) "don't want to look stupid," or (3) "are shy."

If you don't want to be annoying—great! Instructors don't want you to be annoying, either. And your caution about asking questions is not foolish, because although instructors often say, "All questions are welcome!" this statement is dishonest. Annoying questions are not welcome, and some questions are annoying. You'll be less reluctant to ask questions if you know which ones they are.

Questions people ask just to show off are annoying. "Mr. Willingham, don't you think what you've been saying about the history of nineteenth-century Europe relates to the anatomy of tree shrews, which by the way I've been reading about?" No, what I've been saying doesn't relate, and you only asked because you have something *you* want to say about them, and everyone knows it. Don't use my lecture as a platform to show off what you know, with a "question" as your cover.

Questions that sidetrack the speaker shouldn't be annoying but do bother some people. "Mr. Willingham, don't you think what you've been saying about the history of nineteenth-century Europe could be related to the imminent collapse of the aristocracy?" Unlike the tree-shrew business, this question makes sense in light of the subject, so the listener probably isn't just trying to show off. But it will make a few listeners roll their eyes, and I understand why. They are thinking, "You're taking up time with a topic that the instructor didn't think was important enough to include in the lecture. It's great that you're interested (I guess), but why should we all have to listen as you indulge your enthusiasm?" Most people don't hold this attitude, and they recognize that curiosity should be tolerated (at the least) in a setting in which people aim to learn. But if you're very anxious about annoying a few people, fine, don't ask questions that explore new terrain. Talk to the teacher on your own.

The type of question that never annoys others is the one you're most likely to ask: questions of clarification. You miss a definition, so you ask for it to be repeated, or you know I said there were three reasons something is true, and you got only two of them. Fellow students who did get the information understand that everybody misses things now and again, and to the extent that you're "slowing things down," it's for all of ten seconds.

Now, what if the instructor just spent the last fifteen minutes explaining something complicated—say, the octet rule in chemistry—and you realize you Just. Don't. Get it. Can you ask the instructor to explain everything again? You might worry that everyone else must have understood, so asking for clarification will make you look stupid. It's different from the I-missed-what-you-just-said question because it requires understanding. You're not saying, "I didn't hear that," you're saying, "I heard it, but it didn't penetrate my thick skull." Furthermore, the explanation was long, so concern over wasting time is not unreasonable.

The way you phrase your question can alleviate some of each concern. Ideally, you won't just say, "Uh, can you explain that again?" You'll start by saying what you *do* understand. That will help the instructor focus the explanation (making it shorter) and has the side benefit of showing everyone you're not hopeless; you understood some of it.

If you are a worrier, that advice might help, but it's probably not enough. To go a little further on this issue, I'll ask you to get out of your own head for a moment and take the teacher's perspective.

When you ask a question, you're not just helping yourself. **Questions provide feedback to the instructor.** A half-decent teacher is always scanning faces, trying to gauge whether people look puzzled, but that goes only so far. Direct feedback is better.

When it comes to wasting class time to reexplain something: that's not really your call. I'm the instructor, and I'll decide whether or not it's a waste of time. In making that decision, I'll weigh factors such as how quickly I can reexplain it, how many people besides you are probably confused, and what else I need to cover. If I think it's not worth it, I'll say, "I really need to move on, so let's connect afterward on this." Don't take the "blame" for slowing down the group. It's the instructor's decision.

Finally, let me address the "I'm shy" reason for not posing questions. Being ready to ask questions and to admit ignorance is not just a technique for short-term gain in classes; **it's a skill you need to master.** Everyone's job has duties that run counter to their personality or abilities. For example, an extrovert may love that his sales job requires constant contact with new people, but he still has desk work to do in the home office one day a week. If you're shy, you'll still need, on occasion, to speak up and ask questions to make sure you know what's going on. Can you imagine a navy pilot failing to understand a mission briefing and thinking, "I don't want to ask a question and look stupid. I'm sure I'll figure it out when I'm in the air"?

So if you don't like to ask questions, don't view that as "part of your personality" and therefore unchangeable. View it as a skill like any other and one you need to work to improve. If you can, sit in the front row so you can't see everyone else; you might feel less self-conscious. Try asking a *short* clarification question about a definition, just for the practice. If you're reluctant to raise your hand and you have a relationship with the teacher, maybe tell her you're working on this skill; she may become more sensitive to times when you're trying to break in. Asking questions may never feel 100 percent comfortable to you, but the more you make the effort, the easier it will become.

> *In a sentence:* Know which types of questions are annoying and which aren't, and if asking the harmless type of question still makes you anxious, view it as a skill you should master.

For Instructors

How can an instructor help listeners understand the high-level connections that they often miss? Obviously, you should make these connections easy to appreciate by making the organization of your talk explicit.

I find that the simplest method is a preview of the lecture—a slide with a list of the topics that I'll cover, corresponding to the second level of the hierarchy that I mentioned. I spend thirty seconds reviewing it, and then each time I move to a new topic, I return to the slide to show where we are. There's research showing that verbal signals help, too, with or without a slide of the outline. Start by telling your listeners the organization to come, for example, "There are five ways that the consolidation of media companies has affected Hollywood." Then begin your discussion of each by referring back to this organization, e.g., "The third way that the consolidation of media companies has affected Hollywood . . ."

Now, what about this listening-takes-work business? People set a low bar for thinking they understand, so they need your help in knowing whether they've really done so. You can use clicker questions that test what you've just taught, but students find such comprehension questions irritating, and they don't encourage deeper thinking. I prefer to pose a discussion question that requires using the new concept and having students turn to their neighbors and talk about it for thirty seconds. This makes it obvious to students whether they understand a concept well enough to use it.

But recognizing that they don't understand may not be enough to get them to ask a question. They need to feel comfortable doing so, and your body language and facial expression are important cues to your openness. Try videoing a lecture and watch yourself with the sound off, focusing on moments you ask for questions. Do your face and body show openness, eagerness? If you can't tell, ask someone else.

Your reaction to questions is a key determinant of class atmosphere, and the best test case is when a questioner makes it obvious that he wasn't listening. If you shame the questioner, even obliquely, everyone else gets the message: there *are* stupid questions, and those who ask them will pay. Just answer the question at face value and briskly move on.

Even more, look for opportunities to praise questions. Actually, I more often praise the thought that went into the question, rather than the question itself, by saying something like "Oh, that's an interesting insight" to acknowledge that the question had some thought behind it. And there's nothing wrong with pausing after a question to show that you're thinking about it, taking it seriously.

A final note: If your students consistently do *not* ask questions, you should wonder about your relationship with them. They are not quiet because your explanations are so brilliant and clear. They're quiet because they see asking a question as taking a risk. Ask yourself why that is.

Summary for Instructors

- Start a lecture with a visual preview of the organization.
- Return to this preview as you transition to a new topic.
- Reinforce this visual cue about the transition with a verbal cue.

- To help listeners evaluate whether they are understanding, pose questions that require people to *use* the information they have just heard.
- Encourage questions by showing through your facial expression and body language that questions really are welcome.
- When appropriate, praise questions.

CHAPTER 2

How to Take Lecture Notes

Understanding a lecture is hard, and taking notes obviously makes that hard job more difficult—it's an added task. It's no surprise that people don't do it very well. Research shows that if a lecturer lists the points she thinks are important enough to make it into listeners' notes and then examines the notes listeners actually took, she'll see that they captured between 25 and 50 percent of them. That figure doesn't change from middle school through college.

It's not that people are lazy or stupid. Taking perfect notes is literally impossible, because lectures move too quickly. People can speak about six times as quickly as they can write (120 versus 20 words per minute). **Taking good notes requires making wise compromises.**

Items 1 and 2 in the list below describe the mental processes required to understand a lecture. Items 3 through 7 describe the added mental processes required when you take notes.

Mental Processes Required to Attend a Lecture

1. Resist distractions and maintain your attention on the lecture.
2. Listen and understand. The content is probably new to you and complex.

Mental Processes Needed to Take Notes

3. Evaluate the content for importance so you can decide what to include in your notes and what to omit.
4. Decide how to paraphrase the ideas in the lecture.
5. Physically write or type your notes.
6. Shift your gaze between your notebook (or laptop) and the instructor.
7. Coordinate all the processes listed above and shift your attention among them. In other words, decide when to do each of these mental processes and for how long.

The list makes it obvious that taking notes while listening is sort of like playing chess, watching a mystery movie, and cooking a stir-fry all at once. There's not enough attention to go around, so one or more mental processes are going to get shortchanged.

Which processes get cheated of the attention they need? Typically, the first thing you try to do is write or type faster (process 5 in the list). Your handwriting gets a little sloppier or you make more typing errors, but who cares? It's a small price to pay if you can keep up.

But you can't. So what's next to go? Paraphrasing ideas (process 4) is pretty attention demanding, so when people feel rushed, they start to take notes using the precise words the instructor says; that way they don't have to think of a way to rephrase them. But if they're focused on recording snippets of what the instructor says, they can easily drift into a more shallow understanding of what she's saying (process 2). That shallow understanding means they can't properly evaluate what they should write and what they need not (process 3).

So the three mental processes that you will likely shortchange in an effort to write faster all concern understanding: understanding ideas, evaluating their importance, and paraphrasing them. As you feel the pressure of falling behind in note taking, your brain drifts toward writing more and understanding less. You may even tell yourself, "I know I'm not really understanding all of this, but I'm at least writing it, so I can figure it out later."

You probably think I'm going to say, "Don't do that! Write less and understand more." Actually, it's a little more complicated than that.

How much attention you should spend on understanding versus writing depends on the content of the lecture and your learning goals. Sometimes learning means getting a lot of details straight. For example, suppose it's the beginning of your physics lab section, and the teaching assistant is going over the details of how to perform the experiment for the day. There are a lot of details, but none of them are complicated. In this case you'd want to emphasize speed in your note taking, and you wouldn't need to worry much about devoting mental resources to understanding because that part is easy.

Contrast that with a different note-taking scenario: You're a high schooler, and your history teacher tells the class that students can earn a bit of extra credit if they attend an evening lecture on the Great Migration, the movement of about 6 million Black Americans from the rural South to the urban North from World War I until about 1970. To get the credit, students later need to tell the class three important things they learned from the lecture. In this case, emphasizing speed in your note taking doesn't make sense. You need to devote most of your mental resources to listening, understanding, and evaluating the importance of what you hear so you can select your three important ideas.

> ## WHEN TAKING NOTES DURING A LECTURE
>
> **What your brain will do:** It will devote more and more attention to writing quickly in a desperate bid to keep up with the speaker. Little attention will be left for understanding the meaning of the lecture.
>
> **How to outsmart your brain:** Be strategic about balancing your attention to writing and your attention to understanding. The right strategy depends on the content of the lecture; decide on it in advance, if possible.

This chapter will show you how to find that balance, as well as give you some other tricks to ensure that as much of your attention as possible is available to be devoted to note taking.

TIP 6

Be Ready

The main obstacle to taking good notes is time. You're trying to do several things at once. Anything that can be done before the lecture instead of during the lecture should be done before. Prepare.

Make sure you come with the materials you need. **Bring a pen and two spares:** one for you and one for the person sitting near you who didn't bring a spare. Don't use pencils; they are erasable, but they smudge. If you're taking notes on a laptop or tablet, make sure it's charged.

Organize your materials. If you use a laptop, have a separate digital folder for notes. Scan paper handouts into electronic versions so everything is in the same place. If you take notes longhand, buy a separate notebook for each class, and make sure they have pockets for handouts. If you're going to be assigned practice problems or lab exercises, keep those separate so your lecture notes are sequential. Some people prefer three-ring binders, because they make it easy to move pages around, but they are a little heavier. Some students like to have one three-ring binder for all their classes, because that way they never grab the wrong notebook.

If you're the type of person who never seems to have what you need at a lecture, make a list. Get into the habit of thinking each night, "Am I attending a lecture tomorrow?" Tie this question to something you do each night, for example, charging your phone. If the answer is yes, gather what you'll need for the lecture the next day. If you're the kind of person who will gather all this stuff and then forget it at home, put it by your front door so you'll see it when you leave the house.

Many videos about note taking on YouTube encourage you to bring highlighters or sticky notes to a lecture. The idea is to use a red pen for definitions and a blue highlighter for explanations. This sort of thing doesn't make your notes much more useful, and **it takes time and attention to switch ink colors** or put sticky notes onto the middle of a page. It's not worth it.

Though color coding is unnecessary, certainly notes are more useful if you **keep them tidy.** Write the date and subject at the top of the page. Use a broad margin, left and right, so you can add information later. If you're taking notes on a laptop, use a new file for each day of notes. Name the file with the date in this order: year-month-day (e.g., 22-03-18). That way the computer will order the files in the folder chronologically. (You can't count on organizing the folder by date created, because you will update the files later.) Add information about the topic to the file name later. For the love of all that is holy, use a sync program that backs up those files automatically.

If possible, arrive at the lecture at least five minutes early. That gives you a chance to catch your breath, get out your stuff, and turn off your phone. Even better, you can glance at any documents you were assigned to read before the lecture (or glance at your notes from the last lecture) to get your head into the topic.

These may feel like minor details, but they are all directed toward the goal of saving you from having to think about things other than the speaker during the lecture. And these small details can add up to significant distractions if you don't attend to them.

> *In a sentence:* Attention is scarce during a presentation, so minimize the need to perform unnecessary tasks.

TIP 7

Determine in Advance Whether You Plan to Understand More or Write More

I've emphasized that as understanding demands more attention, the amount of information you can record in your notes will drop. Therefore, **think about what you hope to learn**, and consider what other resources are available to support you.

Let's consider two examples representing opposite ends of the spectrum. Imagine that you're a college student taking a creative writing class. Each week, three people submit about ten pages of fiction they've written for the rest of the class to read. In class, roughly twenty minutes is devoted to a discussion and evaluation of each person's work.

Now imagine that you are a high school student taking an American government class. You have an assignment: to write a ten-page paper in which you find a quotation from one of the founders and contrast the quote with principles in the US Constitution. None of the students understand the assignment very well, so the teacher is elaborating on it in class by offering examples of the kind of quotations he had in mind, who counts as a "founder," and what he means by "principles" in the Constitution.

Both of these scenarios call for learning by listening, and you'd want to take notes in either case, but their demands are quite different. In the writing class, your notes would be infrequent and personal to you; different people in the class would get different insights from the discussion that they might choose to record. In the civics class, you'd want to list *all* the details in your notes and get them right; a mistake could create a lot of extra work for you later.

So you want to **consider the relative importance of understanding and of capturing details before you sit down to listen and take notes**. Most of the time, understanding is going to be more important than capturing information, because the details are recorded somewhere else; you can get the facts from a book, and the point of a lecture is to have a live human provide a good explanation of their meaning. But if you find yourself in a parade-of-facts lecture that probably means you *can't* get the information elsewhere, you'd better plan to write fast.

If you want to emphasize getting as much information into your notes as possible, your strategy is straightforward: write as fast as you can, and don't worry too much about deep understanding or phrasing things in your own words. That said, **never write anything that you don't understand**. You may think to yourself, "Not totally sure what she means by 'Technology innovations are usually like a pie shell with half the filling gone,' but I'll figure it out later or ask someone." It's not going to make any more sense later than it does now. And if you ask someone, "What did she mean with that pie thing?" the odds are good they'll say, "I don't remember that." Ask the speaker for clarification immediately if you can (see tip 5) or make a note to ask later (see tip 11).

What if you're thinking you need to focus on understanding? You still want to write quickly, but you need to avoid slipping into using the instructor's words. The easiest strategy is to **understand what the speaker is saying, then write what you're thinking, not what the speaker said**. That will ensure that you pay attention to meaning, and it can also save time. Suppose the instructor says, "Basically, in light of the fact that President Bush was completely and utterly exhausted by his campaign for reelection, there was an expectation on the part of his cabinet . . . or, I don't know, maybe not an expectation, maybe more of a fear . . . anyway, they thought that maybe the first quarter of his new term would be wasted and the so-called honeymoon period was just going to pass by before his energy returned." You should write: "Campaign exhausted Bush; cabinet worried he'd rest, waste political capital."

Paraphrasing actually has another benefit: it helps memory, for reasons I'll explain in chapter 3. For now, I'll ask you to take my word for it.

In a sentence: If a lecture is detail heavy but easy to understand, focus on recording as much as you can; if the important content is more abstract, focus on understanding and write notes sparingly, using your own words.

TIP 8

You Should Usually Take Notes Longhand

Should you take notes with paper and pen or with a laptop? First, note that this question assumes that you have a choice. Devices are sometimes forbidden and sometimes required in class, and in some settings a device just doesn't make sense—for example, if the lecture has a lot of figures that would be hard to capture with a device. If you do have a choice, you should again **consider the relative importance of understanding versus recording a lot** in your notes.

Let's start with speed. With some experience, people can type faster than they can write. That seems like a pretty important advantage, given that we keep coming back to speed as a key problem. But typing quickly can tempt you to try to record everything because it seems more possible. One experiment (widely reported in the news) showed exactly that: people taking notes with a laptop were more likely than longhand note takers to write snippets of what the instructor said word for word. But other studies have not found that effect, so it's unclear how general it is. Basically, laptops have the edge over longhand if you're very concerned about getting a lot of the lecture into your notes, especially if you can fight against slipping into dictation mode.

But that possible advantage might be canceled out (or worse) by the disadvantage of distraction. If you've got a laptop open, your email, social media, shopping, and other distractions are a click away. It is very difficult to resist the impulse to have a little look at the internet, and you're a fool if you make that distraction easy for yourself. Someone who attends a lecture and brings a device with internet access, saying to himself, "I'm just going to take notes," is like an alcoholic who goes to a bar, swearing he'll have just a few appetizers. A very general, very wise principle of human behavior is: ***Do Not Rely on Willpower if You Can Change the Environment Instead.***

As a professor, I would happily change the environment for my students—if I could—by turning off Wi-Fi access in my classroom. I asked the IT group at the University of Virginia about the possibility, but they pointed out that even if I switched off the router in my classroom, the whole campus is saturated in Wi-Fi; students would just pick up a signal from another router.

An alternative is to **put your laptop into airplane mode**. That way it's a little harder to access online fun, and you're more likely to stay with the presentation.

Another problem is that your use of a laptop may distract others. Certainly, some of my students complain about this. Human biology works against us; when you perceive something moving in your peripheral vision, your brain is wired to direct attention to the movement. We can easily see the evolutionary significance. For our distant forebears, something moving might be a threat, and it had better be checked out right away. Now, eons later, a fellow lecture attendee flips through pictures of shoes on Zappos.com and the lizard part of your brain shrieks, "WHAT'S THAT!?!"

That's the concern, and my students do feel it's an issue, but proving that it's a problem in formal experiments has been harder. Again, there's one study reporting that students watching a videoed lecture (on which they were to take notes) were distracted by someone surfing the web in front of them. That study got a lot of press, but other researchers have failed to find the same effect, so it remains unclear just how big a problem it is.

What I've described so far is a sort of microanalysis of what might or might not happen if you use a laptop to take notes. Wouldn't it be simpler for researchers just to compare how much people learn if they take notes on a laptop or by hand? An experiment based on real life would compare the final grades of college students who use laptops to the grades of people who take notes with paper and pen. But *that's* an imperfect method, too. Maybe people who choose to use laptops are generally less motivated to get good grades. Or maybe easily distractible people enjoy using laptops more; who knows?

All in all, research does not provide a clear answer to the laptop/longhand notes debate. My experience as a college instructor tells me that being lured to other online activities is a serious problem. In anonymous surveys my students say my lectures are interesting, yet when a colleague observed my class a few years ago, he told me that *many* of my students using laptops had been off task. That prompted me to pop in on other lecture classes at my university, and I saw that the problem was extremely common.

So what's the bottom line? **If the lecture is not a fact-heavy one where speed is essential, it's best to take notes longhand. If speed is essential, use a laptop but disable your Wi-Fi before the lecture starts.** And if you find yourself distracted anyway, switch to longhand.

In a sentence: Although the research on laptop use during presentations is inconclusive, I think the presence of the internet is so distracting that you'd be wise to take notes longhand in most circumstances.

TIP 9

Evaluate Your Notes on the Spot

I write a lot of notes to myself, and not just in classroom settings. I get ideas at odd times, and I learned in college that I quickly forget them even though they seem like brilliant revelations at the time. That was long before smartphones, so I got into the habit of carrying a small pad of paper and a pencil stub, and, yeah, people thought that was a little weird, but people already thought I was exactly the kind of guy who would carry around a notepad and pencil stub, so it didn't make that big a difference.

But it turned out that the notepad didn't solve the problem, because I was still too confident that I would remember my great insights; the notes I took were too short. I remember having an inspiration for the introduction to my senior thesis when I was hiking with friends. I scribbled a note and later saw that I had written, "Don't forget about the finger puppet." I spent a lot of time over the next few days trying to see some connection between my thesis and a finger puppet. I never got there.

I've said that it often makes sense to "write what you're thinking" (see tip 7), but you must bear in mind that *future you* will be reading the note. **Write your notes for future you.** Future you needs context and explanation, which are not easy to provide when you are rushed during a lecture. Also, you don't want to go too far in the other direction, providing details when they are not needed. I remember for one of my college classes I bought a used copy of a book of John Keats's poetry, and it was full of the previous owner's notes. One line from "Ode to a Nightingale" reads "Thou wast not born for death, immortal Bird!" The last two words were circled, and written next to them was this observation: "Bird, you are immortal." This kind of note taking may not be the best use of your time.

So how do you know if your notes strike the right balance between briefness and clarity? **When the instructor calls for questions, evaluate whether your notes will make sense to you later.** As I said in chapter 1, that's a good time to check your understanding: Do you see how the many facts in the lecture relate to one another and build to a larger point? You should also check your notes: Do they capture your understanding? At the very least, check for incomplete thoughts, abbreviations that make no sense, and graphs with an unlabeled axis. But look, too, for finger-puppet notes, references that might seem clear now but might not in a few days.

I've emphasized that you have two jobs when you're learning by listening: to understand in the moment and to take notes so that you will later have cues to prompt your memory. When speakers pause, they usually refer only to the first task. They ask, "Any questions?" by which they mean, "Do you understand?" They don't ask, "How are your notes?" Check them anyway.

You can also take a few moments to **evaluate your notes at the end of the lecture** if you don't have to run to another appointment. This is a great time to look for holes in your notes, because the lecture is still fresh in your mind, and if a question does come up, the instructor may still be there, available to answer it.

In a sentence: Evaluate your notes as you take them, to see if they will make sense to you later.

TIP 10

Don't Use a Note-Taking System

They are often called note-taking systems, but a better term is probably "note-taking formats": they describe how to lay out ideas. For example, the mind-mapping technique has you write notes as a kind of web. Instead of writing sentences or phrases on lined paper (as in traditional notes), each entry is just a word or two. You write the main topic at the center of a blank page. Then ideas radiate outward.

There is some experimental evidence that note-taking systems help. High school and college students take better notes and get better grades when they use a system, whether it's mind mapping, Cornell, the charting method, or one of the others. Still, there's no evidence that one particular system is more effective than others, because the experiments almost never compare one note-taking system to another. They examine whether offering instruction in a note-taking system is better than offering no instruction on how to take notes. A likely explanation for this pattern of results is that most people take pretty bad notes and almost anything you do to get them to think more deeply about the process will help.

I don't recommend that you use a note-taking system, because I don't think they are worth the cost to attention. Using a special format for notes is just one more thing for you to think about when your mental state is in near overload most of the time.

Rather than adopting a formal system, I advocate writing notes more or less the way you're used to doing it. That way you don't have to think about it and can devote more attention to understanding the lecture. **Use phrases and broken sentences you can understand.** If it helps, imagine that you're texting someone. For organization, use a minimalist outline format that you feel comfortable with. When I was a student, I used three levels of headings: capital letters, numbers, and then dashes.

In chapter 4, I will show you how to reorganize your notes into better written form later. To facilitate that process, I recommend that you **take notes on alternate pages**; in other words, leave every other page blank. You'll use the blank page to amplify your notes and reorganize them (if need be). I recommend writing your first set of notes on the left-hand page within a spread, because in this culture we write from left to right; it's going to seem more natural that the notes (which came first) are on the left and your notes on the notes appear on the right. It's a small detail but worth it. (Naturally, if you're taking notes on a laptop, editing will be easy, so this isn't an issue.)

The only note-formatting trick I advocate is learning some short-hand abbreviations. That's the subject of tip 11.

> *In a sentence:* Don't use a special note-taking format, but leave plenty of room for future editing and annotations by writing on every other page.

TIP 11

Use Note-Taking Shorthand

As we've seen, speed matters when you're taking notes, and using some easily learned abbreviations will help. I offer some suggestions in the list below, but there's nothing magical or research based about them. If you find or invent others you like more, use them. If you have abbreviations you're familiar with from texting, use them. Also, I don't recommend trying to memorize a large set of abbreviations and then agonizing about their use while you're taking notes. That defeats the purpose. Add one or two per week or at whatever time interval feels comfortable.

Because: bc
Years: yrs
With: w/
Without w/o
Within: w/i
Amount: amt
Something: s/t
Somewhere: s/w
Someone: s/o
Important: imp
Minimum: min
Maximum: max
Versus: vs
Between: btw
Example: ex or e.g.
Before: b4
Equal, equivalent, the same: =
About the same: ≈
Not equal, different: ≠
Greater than, more, bigger: >
Less than, less, smaller: <
Increase, growing, improving: ↗
Decrease, shrinking, getting worse: ↘
Leads to, creates: →
Change: Δ
Again, repeat: ↻
None, never, not: ∅

And: &
Regarding: re
Though: tho
Compare with: cf
Number: #
That is: i.e.
Quarterly: ¼ly
Annual: ann

Look for frequently used words, and abbreviate them with a single letter. In a psychology class, *S* stands for "subjects," in an education class *S* means "students," and in a chemistry class it means "sulfur." In a course on ancient civilizations you may spend just a day on the cultures of third-century-BCE Mesopotamia; *M* ought to have a particular meaning that day.

Occasionally it's crucial to record the exact wording of something the instructor says, or you come up with a good paraphrase you want to write. In both cases, your mind has a fairly long string of text to write, and it's easy to forget the ending of that string as you're writing the first part. That's especially true because the instructor is still talking. One trick that can help: write the first letter of each word of what you want to write, leaving space to fill in the rest later. So if you hear:

"Four score and seven years ago our fathers brought forth, upon this continent, a new nation, conceived in liberty, and dedicated to the proposition that 'all men are created equal,'"

you write:

4 s	& 7 yrs a	our f	b	4th u	this c	a
new n	c	in l	& d		to the p	
that all men are c		=.				

Then you go back and fill in the words, given the first-letter clues you've created.

Figures and graphs can send you scrambling: they are complex and take a while to draw. You might consider taking a picture with your phone, but that takes time, and it's often frowned upon in classes. If you need to copy a figure into your notes, **be sure you know what the *point* of the figure is and write that conclusion in words.** Have a look at this figure:

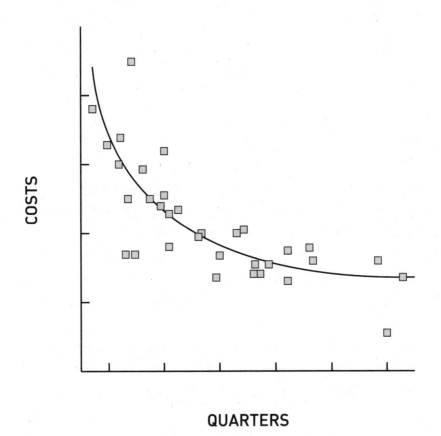

A speaker might show this to make one of the following points (or others):

1. This company is doing a terrific job in reducing costs.
2. This company initially did a terrific job in reducing costs, but the reductions have slowed.
3. This company probably cannot look forward to future reductions in costs and so must find other ways to increase profits.

If you are clear on the conclusion, you can draw the graph in a way that makes that plain and also makes it obvious whether you need to label axes or part of an axis. For example, if the speaker wants to highlight that costs have dropped by $6 million, bracket the drop and label it "$6 mil." If the speaker emphasizes that cost savings slowed during the recession of 2008, mark that spot on the horizontal axis. And so on.

Although I don't recommend using a formal note-taking system, it can be very useful to add comments to your notes that will help organize them later. **Use the page margin for notes to yourself about your notes.** Here are a few ideas for the types of notes you might want to add, with shorthand symbols you might use.

I missed something here: ?
I'm confused/missed the whole point: ??
I understood this bit but not what it connects to: ?→
I'm writing this, but I'm not sure it's right: OK?
Key conclusion/important: *
This is my idea, not the instructor's: (M)
Now my notes refer to the speaker again: (S)
I think this is a digression: (D)

The point of shorthand is that it will help you record more of a fast-moving lecture. But what if the lecture is recorded? That's the subject of the next tip.

> *In a sentence:* Reduce the mental burden of note taking by using your own abbreviations.

TIP 12

Use Lecture Recordings Judiciously

Some speakers provide listeners with an outline of their talk or their slide deck. These resources provide both an opportunity and a danger. I pointed out the danger in chapter 1 (see tip 3): you will likely pay less attention during the lecture if you figure you can always review the written material the instructor has provided.

What if a video recording will be available later? Or say you are permitted to record the audio. Since you can review the lecture later, doesn't that remove most of the pressure to write quickly?

You may fully intend to revisit a lecture later so you can supplement your notes, but you probably won't. **Watching a video or listening to audio is like attending a lecture again, and that's a big investment of time.**

It sounds as though I'm saying I'm pretty sure that you're lazy. Most of us are! Well, maybe not lazy but certainly busy. I've made a habit of asking my students whether they listen to the lectures they record. They mostly don't, and the reason they offer makes perfect sense: they imagine that they will at least use the recording as a backup—they will revisit the parts of the lecture that they didn't quite understand—but they later realize that it's much easier to ask a classmate (or me) for clarification. It's hard to find the relevant part of the lecture on the recording, and when they do, they often don't understand it any better upon a second listening. They need a different explanation, example, or analogy.

The availability of a recording may tempt you to skip a lecture if you're busy with other things. A lot of research has been conducted in the last ten years that compares the effectiveness of learning from a live instructor versus learning from video. Much of the research is poor, but what we have indicates that **a live lecture has the advantage**. We can imagine a few reasons this might be so. Although the idea of watching a lecture in your pajamas sounds appealing, your attention is also probably more apt to drift away if you're watching a video at home. The video drones on while you sneak out to the kitchen to grab a snack, or you figure you can just listen to the audio while you explore Reddit on another tab. Then, too, you can't pose questions to a video, nor can you benefit from the questions that others pose. Yes, I know, on some platforms you *can* ask questions and you can see other people's questions and the answers. But sometimes it's troublesome to access them, or someone poses a question and the instructor hasn't gotten around to answering yet, so you have to remember to check back. As noted above, we're all a little busier than we think.

So the upshot is that you should **think of a lecture recording as an emergency backup** or as an insurance policy. You should hope you won't have to use it, because it's an inconvenient and inferior substitute.

> *In a sentence:* Video or audio recordings lull you into thinking that you don't need to fret about capturing a lecture in your notes, but don't fall for that; you're less likely to use the recording than you think.

For Instructors

I've emphasized that the disconnect between listening and writing is one of speed. The most obvious advice for instructors is to speak more slowly. It's worth asking someone about the pace of your lectures; it's hard to judge yourself and doubly so if you try to evaluate your pace at the same time that your mind is occupied by lecturing.

You can also help students by telling them what they can omit from their notes and what's essential. For the latter, *stop talking* and give them time to record what you've said. Likewise, if you want them to check over their notes to see if they are complete and comprehensible, periodically give them time for that.

Even if you don't provide copies of your slides to students, consider providing copies of complex figures so listeners won't spend time frantically copying them. Put a distinctive mark (say, a red dot) on each figure that you'll provide so they'll know that there's no need to copy it.

On the subject of slides, remember that listeners will likely write whatever is on a slide, probably verbatim. They use "presence on a slide" as a rough measure of importance. Generally, that should mean that you put less text on slides in an effort to get your listeners thinking more and copying less. But if you want them to write something word for word—a definition, for example—put it on a slide.

I've suggested that learners take notes longhand, based on my own experience that the availability of internet fun close at hand is too great a temptation for most people. Should you simplify matters by making the choice for them by banning devices? Here are a few thoughts that may help with this difficult decision.

First, I do think it's useful for you to set the policy. Some instructors permit laptops, stipulating that they can be used only for learning-relevant tasks. But that makes the instructor responsible for monitoring that the rule is followed, which distracts everyone.

Second, consider whether your lectures tend to be fact heavy and fast paced, in which case learners may benefit from the speed that typing affords, or whether your course tends to be slower paced and contemplative.

Third, ask your students what they think about the matter. I find I get more thoughtful responses when I pose this question at the end of the semester—that is, I ask them what my policy should be in the future; it's easier to choose a wise policy for someone else rather than oneself.

Fourth, if you ban laptops, remember that some students use them to address a motor control problem or other disability. Avoid possible embarrassment by announcing that any student who strongly prefers to use a laptop for any reason can have a conversation with you about the matter.

Finally, what about lecture notes? Again, I encourage you to consider the trade-off between the two functions of note taking. Providing an outline ensures that students will have a complete reminder of the ideas you presented, but some will likely tune out, thinking, "Eh, I've got an outline." Providing a skeletal outline offers a compromise. That shows listeners the high-level structure of the lecture, which should be a significant help in understanding the main points and their organization, but it is not so complete that listeners will be tempted to daydream.

Summary for Instructors

- Talk more slowly.
- Signal when something should be written in listeners' notes, and then pause to allow them time to write it down.
- Distribute copies of figures and visuals, and let listeners know which ones they don't need to copy.
- Bear in mind that students copy what's on slides, whether doing so makes sense or not.
- Forbidding learners' use of laptops may make sense in some circumstances, but there are many factors to consider, including the norms of the institution, the attitudes of the learners, the information in the lecture, and what learners will be expected to do with that information.

CHAPTER 3

How to Learn from Labs, Activities, and Demonstrations

Instructors tend to do a lot of talking because it's an efficient way to communicate new information. But good presenters know that learners can listen for only so long. No matter how good the lecture, after a while listeners feel the need to move around a little or do some talking themselves. So a good speaker intersperses other activities into a lecture—a demonstration, perhaps, or small-group discussions. In other situations, the instructor does very little talking, and most of the session is learning by doing, as in a high school biology lab.

When someone lectures, it's pretty obvious what you're supposed to learn. People lecture to convey information: facts and how to do things. But activities can serve different learning purposes. That matters, because you should adopt different learning strategies depending on what you're meant to learn. In this chapter I'll show you how to learn from different types of activities.

Let's start by considering the three main purposes of learning activities.

The Purposes of Learning Activities

First, some activities are meant to teach you a **process**—that is, how to do something better. Aristotle had this purpose in mind when he said, "Men become builders by building and lyre players by playing the lyre." Process is what you're to learn during a guitar lesson or when you're being taught how to affix and slice a brain in a biology laboratory.

Second, you might do an activity for the **experience**, because doing is the best or only way to learn particular things. I can *tell* you that the enormous height of cathedrals makes worshippers feel awed, but my description can't possibly evoke the same feeling as standing in that huge, hushed space. Some things—for example, what it's like to work in hospice care—can't be learned via a lecture or a book; you must experience them.

Finally, sometimes the doing is meant to help you **understand** something, especially when the thing you're to learn is difficult to put into words. For example, when teaching about circles, an elementary school teacher might have students form a line on the playground, pick a student at one end to be the "origin," and have the line of students walk around the origin in concentric rings. Students can memorize the formula for the circumference of a circle ($2\pi r$), but they get a better understanding of why the formula works if they see the kids near the origin taking tiny steps, while those far from the origin must run—as r increases, the circumference of the circle increases.

Now you can see the problem in learning from activities: the purpose is often not obvious to students. When students create the circumference of a circle on a playground, they might understand that this activity is meant to help them understand $2\pi r$. But they also might think that the goal is the process—an exercise in teamwork and cooperation. Or even that the purpose is the experience itself—getting outside in the sunshine and moving their bodies.

In addition to understanding what's to be learned, knowing where to direct your attention is also a problem when you participate in activities. Naturally, if you don't pay attention, you can't learn; if a student doing the circle-on-the-playground activity is thinking about whether the grass is making his shoes dirty, he'll miss the point. But the relationship of attention and learning is even more subtle than that, and we need to clarify it to ensure that you know how to learn from activities.

How Learning Follows Attention

Naturally, if you're not paying attention when there's something to be learned, you won't learn it. But even if you're paying attention, you usually can't pay attention to *all aspects* of the thing, so you will later remember only that part you did pay attention to.

Here's an example. Suppose a new family moves in next door, and I think, "I'll take them a basket of stuff to welcome them to the neighborhood." I want to put a pound of nice coffee in the basket, so I plan to buy it on my way home from work. That evening, when I drive by the grocery store, I think to myself, "Do I need anything from the market?" I conclude I don't and drive on. As I pull into my street, I see my new neighbor checking her mail, and I immediately think, "Darn it! I forgot the coffee!"

How come the sight of my neighbor prompted me to remember but the sight of the grocery store didn't, even though that's where I usually buy coffee? This sort of thing can happen when there's a mismatch between the way I search for a memory and the way the memory went into the vault.

When I run out of something such as coffee, I think about it as a deficit; there's a list of staples I should always have in the house, and when one is missing, I make a mental note to make up the deficit. That's different from thinking about coffee as part of a welcome gift to new neighbors. I probed my memory by asking, "Do I have any grocery deficits?" But that morning I hadn't thought about coffee in the grocery-deficit sense; I had thought about it in the neighbor-gift sense.

The particular way that we think about things is a key contributor to what we remember. It would seem that if you think about a chair, later you'll remember having thought about a chair. That sounds obviously true, but you've just seen that it isn't. You can think of a chair as something to sit on, as something with glued joints, as a sign of status if it's at the head of a table, or as a weapon in a barroom brawl. *How* you think about it determines what you'll remember later. Elsewhere I've described the idea this way: **"Memory is the residue of thought."**

This principle is especially important when we consider learning during activities. Something as simple as coffee or a chair has lots of different features, and you've seen that you'll remember the features that you pay attention to and won't remember the others. So selecting the right features of an activity to pay attention to is essential because it determines what you'll learn from it.

Attention, Memory, and Learning by Doing

Where you should direct your attention is pretty clear when you're learning by listening. You should focus on the speaker, and she will tell you whether she wants you to think of a chair as something with glued joints, something that can replace a step stool, or whatever.

When you learn by doing, it would be ideal for the instructor to provide the same sort of direction: "I want you to do *this*, and while you're doing it, I want you to notice *that*." But a lot of times instructors don't provide the guidance because they don't understand that it would help. If you know something, it's hard to conceive that others can't easily see it or figure it out. This problem is commonly called "the curse of knowledge." If you've ever played charades, you've experienced it. To you, your pantomimed actions are so *obviously* someone preparing breakfast that it doesn't occur to you that they are actually compatible with another interpretation, such as performing surgery.

When a teacher asks third graders to make music by lightly striking glasses holding varying amounts of water, it seems obvious to her that the pitch produced relates to each glass's water level. Thus, she thinks the activity needs no preparatory remarks, no setup—in fact, that telling students what to expect will diminish its power, like explaining a joke.

WHEN LEARNING FROM ACTIVITIES

What your brain will do: It will store in memory whatever you direct attention to and fail to store whatever you don't pay attention to; when you learn by doing, there's more than one possible target of attention.

How to outsmart your brain: Decide, as strategically as you can, where you will direct your attention before the activity begins.

Where does that leave you? If you're told where to direct your attention during an activity, great. If not, you must make your best guess. This chapter tells you how to guess wisely.

TIP 13

_____

Be There and Engage

If you're to learn from an activity, **you need to actually participate**. If you're in a private guitar lesson, you can't get away with doing nothing. But if the "doing" is a group discussion or a field trip, it's pretty easy to shirk. If the instructor thought you could learn the same thing by watching other people do the activity or by reading about it, that's probably what you'd be asked to do, because it's easier for everyone. You're being asked to do the activity because there's no other way to learn. So do the activity.

In the same vein, **do the prep work** expected of you. If you are asked to read something in advance, to bring something to the activity, to try something out, or to practice something you've already done—do it. Aside from the fact that ignoring the instruction means you won't be fully prepared, there's also a chance that you will feel awkward during the activity and you'll mentally withdraw.

These are pretty obvious recommendations. Less obvious but equally important is an implication of the memory research I described: if you're not present, you not only miss the opportunity for understanding, you also miss the chance to take your own notes. **The notes you get from someone else will not be the same as those you take yourself.** Notes are cues that will jog your memory back to the understanding you had during class, and you've seen how particular those cues can be; grocery-deficit coffee is different from neighbor-gift coffee. If you can't make it to an activity but figure you'll ask someone about it or look at the notes they took, you will be getting *their* cues to *their* memories.

"Be there" also means "Be sure you won't be distracted or need to leave in the middle." If you wear glasses, bring them. Have a spare pen. Make sure your laptop is charged. Don't come hungry. If you get cold easily, bring a sweater. Go to the bathroom before the activity. If you're a smoker and will want a cigarette, have one right before you start. Don't leave to take a call or answer a text.

An activity can feel like a break, like a day off from the harder days when you're expected to learn. Don't fall for that feeling. Be there and engage.

> *In a sentence:* Activities make a nice change of pace, but remember, you're there to learn, so come prepared and stay focused.

TIP 14

If the Activity Is Brief and Offers a Surprising, Interesting Experience, It's Probably an Analogy

Sometimes an instructor will have you do something as a way to help you understand an idea, especially one that is difficult to explain with words and pictures alone. Maybe the idea makes more sense when you see it in action or have a chance to explore and manipulate it yourself rather than hear about it. For example, it's not easy to describe the full range of meanings of the equals sign in mathematics. So a teacher might have children use a pan balance with digits, each weighted proportionally to its value. They can see which values make the scale balance, which changes maintain the balance and which do not, and so on. In a physics class, students might experiment with a bicycle wheel mounted with handles on the axle to help them understand angular momentum: What changes as the speed of the wheel increases?

Or consider this example. A high school history teacher asks her students to write their own version of the Declaration of Independence, a letter from the thirteen colonies to King George III. It's to be historically accurate but written in the style of a breakup letter. What's the goal of this activity? The breakup letter is an analogy. An analogy pairs *something you already understand* with *something you are trying to understand*. High school students already understand that a breakup letter has three elements: a couple, a member of the couple who wants to end it, and an explanation by that person to the other. Students know that the colonies and Great Britain had a relationship and that the colonies ended it. But most probably think of the Declaration as a sacred historical document or perhaps as "the thing that started the Revolution." They don't think of its function. Just as a breakup letter communicates "I want it over, and here's why" from one member of a couple to the other, the Declaration communicated "We're unhappy and it's over, Great Britain, and here's why."

The key to learning from an analogy is paying attention to the right features, because every analogy has features that matter and features that don't. When someone says, "Lawyers are sharks," she means that both are tough, ruthless, and frightening. She doesn't mean that lawyers have gills. Likewise, in a breakup letter the author often accepts some blame for the split as a way of softening the blow. That part of the breakup letter analogy doesn't hold; the colonists thought everything was George III's fault.

When the activity is an analogy, **focus on the mapping**. The mapping is the pairing of the features of the *something you already understand* with the features of the *something you are trying to understand*. In this case, it's thinking, "The person breaking up is the thirteen colonies. The person being broken up with is Great Britain. The breakup letter is the Declaration of Independence."

Obviously, **if you are unsure about the mapping, ask**. As I've said, instructors often want you to experience an activity before they explain it. They figure that part of the learning process is your exploring a bit, thinking things over. That's fine, but you don't want the learning episode to end before you've understood the point. If there was reading or some other work you were supposed to do to prepare for the activity, thinking back on that will likely help you figure out the mapping.

Another thing to keep in mind during this sort of learning activity is probably more important: **don't get distracted**. Learning activities are meant to be engaging, but sometimes irrelevant parts of the analogy are really fun to think about. For example, you can imagine that students writing the breakup letter from the thirteen colonies could get really caught up in the format and spend a lot of time thinking of jokes from romantic comedies they've seen. I've heard about a middle school math activity in which the teacher pushes the desks to the edge of the room and uses carpet tape to create a graph grid on the classroom floor. Then the teacher writes a linear equation on the board and the students create a line on the graph; each student represents a point on the line, and they stand on the graph with their hand on the shoulder of the next "dot" to create the line. I can see that the teacher wants them to better visualize the line (compared to drawing it on paper or a computer), but there's the potential for some middle school kids to devote a lot of attention to the hand-on-the-shoulder bit.

Activities are meant to add a bit of zest to your learning experience. Go ahead and be zesty. Just figure out what the activity is supposed to teach you before it's over.

> *In a sentence:* A brief activity that makes you say "Cool!" is probably meant to illustrate some abstract idea you've studied, so make sure you understand how the activity explains the idea.

TIP 15

If the Activity Comes with a Script, You're Supposed to Learn Either Skills or Concepts

Some activities come with a script, a set of steps you're supposed to follow. For example, several years ago the University of Virginia developed a new web-based platform on which faculty would view applications for its graduate program. I attended a half-day training session to learn the new system, and although there was some lecturing, I spent most of the time on my laptop, using the system. But I was not left to poke around and try it out. A sequence of tasks was set for me. I was given a script. Most laboratory courses follow the same game plan: students conduct an experiment, and they are given step-by-step instructions on how to carry it out.

What's the purpose? **One purpose is to make the learning more memorable.** In theory, I could have learned the new application system by memorizing a manual. But enacting the steps is another instance of "Memory is the residue of thought": it's best to put information into memory the same way you plan to take it out. I think the same applies to science labs. Part of the purpose is to learn techniques and methods, such as how to use a potentiometer or how to culture bacteria. You can read about it, but you learn faster if you do it.

Other times, a scripted activity has a different purpose, namely **to teach you higher-level thinking techniques**. You are expected to learn something beyond the specifics of the activity itself. For example, the instructor of a science laboratory may want students to learn something about the scientific method. But expert scientific thinking is complicated. That's why there is a script. If you simply gave people chemicals and equipment and said, "Find out what happens if you try to recrystallize benzoic acid in acetic acid," they wouldn't learn much. You have to walk them through the idea of crafting a hypothesis, creating an experiment to test the hypothesis, and interpreting the results in light of the hypothesis.

So how can you maximize your experience if you are learning with this sort of activity?

First, avoid the typical thinking pitfalls. The most common is to focus on the outcome of the activity rather than the process. It's understandable; you've been given a script, so you think that doing a good job means following the script. If I follow the script correctly, I'll get the expected outcome. So is my benzoic acid recrystallizing? That's the right priority if you're trying to fix your dishwasher by following the steps in a YouTube video. But the real goal of dissecting a frog in biology lab is not to create a well-dissected frog; it's to learn, so focus on the process of what you are doing.

Another pitfall is to think of very little at all. The activity gives you a script to follow, so you mindlessly follow the script, executing instructions but not thinking about why you're doing them.

What *should* you focus on? **The main point of an activity is either to learn a technique that requires physical practice or to engage high-level thinking strategies** such as the scientific method. These two purposes are obviously quite different—one is concerned with details and the other with the big picture—so it's essential to know what you're supposed to get out of it. Again, the obvious thing to do is ask the instructor. (If the instructor says, "Both," you're in the hands of an amateur. You can't think about two complicated things at the same time.) If the instructor refuses to tell you, you can probably get a good idea from the script: if the directions include a great deal of detail about *how* to execute steps, it is a technique activity. If it includes questions and/or directions that could apply to many different tasks, it is a big-picture activity. Either way, let that be your focus as you do it.

In a sentence: If you're given step-by-step instructions, you're probably supposed to learn either the smooth execution of the steps themselves or something very high-level and abstract that the steps illustrate—figure out which one.

TIP 16

For Projects, Pick the Problem with Care, Seek Feedback Along the Way, and Reflect at the End

When you're a student, learning by doing sometimes means solving an open-ended problem—that is, one that doesn't have a single right answer. I don't mean a paper-and-pencil problem that takes half an hour, I mean a project that you work on for weeks and that usually results in a tangible product. For example, a student nearing the end of an accounting course might be told, "Find a small business in town and help it set up a system for its inventory, its tax obligations, or its payroll." I have three suggestions for how to maximize your learning when undertaking this sort of project.

First, **select your project based on what you want to learn, not what you want to accomplish**. It's hard not to be really concrete when you're brainstorming project ideas, because your proposal must be concrete. So you think, "I wonder if I can make a robotic inchworm?" or "I saw a video of Jimmy Kimmel making a rocket out of a Pringles can" or "I'd really like to use the 3D printer." You might commit to a project that results in a cool product but is boring to work on. What if the robotics problem ends up being pretty much snap-together parts? Or you're interested in animal welfare, so you decide to make a video to increase awareness of animal testing in the cosmetics industry—which ends up in your having to spend most of your time learning how to edit videos. So **you want to think about process as you choose your project: Will the process emphasize the elements you are hoping to learn about?**

When you think about your learning goals, remember that you may not need to stick to traditional academic facts and skills. (Naturally, you'll want to consult with the instructor about this.) Maybe you want to learn how to manage your time better, so you pick a project that places frequent and inflexible demands on your schedule, such as creating and caring for a complex aquatic environment. Or maybe you've had trouble working with other people in the past and want to develop your skills as a team member. Craft that goal into your project, perhaps by volunteering at a local charity.

Second, **when you are in the middle of your project, be sure you get feedback**. It's hard to create your own feedback; you may be able to tell that something is going wrong but not know why. Too often my students think they're expected to hand in the completed project at the end, without any guidance along the way. You should seek input from the instructor and from others who might help. The usefulness of feedback is another reason (if you needed one) to stay on schedule; you can't expect to get feedback at the last minute, much less act on it. (I'll have more to say about planning in chapter 10.)

Third, when you've finished the project, **pull your thoughts together and reflect**. Do so in light of the learning goal you set. Did you learn what you had hoped to? Did you learn something you didn't expect to? I urge you to jot down a note or two as you reflect. At the time, it feels as though the lessons learned will stay with you, but don't count on it. The best-case scenario is to use the experience to make your next project a better learning experience; reflecting on it right afterward and recording your reflections may make the difference.

In a sentence: Pick your project based on what you want to learn, not the product you want to produce, get feedback along the way, and take the time to reflect on the process afterward.

TIP 17

When the Purpose Is the Activity Itself, Know the Difference Between Experience and Practice

Sometimes you must *do* to learn because the doing is the thing to be learned; learning by listening or reading won't work. Athletics and the playing of musical instruments are obvious examples, but this category also includes:

- Writing clearly
- Interacting socially
- Being a good team member
- Giving a speech
- Leading a group

Lots of bits and pieces go into these skills, and for that reason they take years to master. They contrast with the simple skills referenced in tip 15, which can be learned in hours—something such as operating a microtome or taking blood pressure.

Aristotle was right in saying that the *doing* is vital—a lyre player must play the lyre—but it's not quite as simple as that. I've *done* plenty of things for decades without improving: driving a car, for example, or baking cakes or typing. How is it possible that I keep doing these things, yet I don't improve? Simple: **experience is not the same thing as practice.**

The reason you do an activity will usually determine what you're thinking about while you're doing it, and that determines whether or not you learn while you do it. I drive my car to get to places, and my friend Adam plays guitar for the pleasure of his friends. But these purposes—to achieve a practical result or to give pleasure—do *not* lead to the improvement of complex skills. When I bake, I'm not trying to get better at baking, because I'm happy enough with the results, so I don't think about trying to improve. And as you now know, what you think about is very important to what you learn.

Psychologists studying complex skills have developed more specific principles than "Think about it" in order to maximize your improvement:

1. You must **focus on one aspect of the skill at a time**. Complex skills have a lot of components, and you can't think about all of them at once. You can't practice "writing well," but you might practice "choosing vivid words" or "varying sentence structure."

2. How should you select the one aspect of the skill to work on? For some skills, there's an accepted order: in learning to play piano, you start with scales and simple time signatures. For skills where there is not an accepted sequence, **start with what seems elementary to you but you're inept at**. Work on

that until you're good at it, then work on the next component of your incompetence.

3. How do you know what you're inept at? It may be obvious—when playing golf, my tee shots often hook—but it may not be obvious *why*. **Feedback is vital,** not just on the outcome but on what you're doing that makes the outcome unsatisfactory. You might get good enough feedback by observing yourself, but you'll probably need someone who's better at the skill to observe you and tell you what you're doing wrong.

4. It's not enough to confirm via feedback "Yup, I'm terrible at that bit." You need to **generate and try out new ways** to do it. Up until now, when you have noticed yourself using the same word repeatedly in an essay, you have hauled out the thesaurus and picked another word. But people have said that those substitute words feel a bit off. Your strategy to improve your word choice isn't working, so now what will you try?

5. You have to **concentrate on what you're doing.** That sounds like throwaway advice, but it's probably the most important difference between intentionally practicing something and simply doing it. Experience enables you to do things without much effort; you've done them so often that you're on autopilot, and it seems as though you're scarcely thinking about the process at all. But when you practice, you focus on one aspect of the skill, try out new methods to do it, and monitor the results. That's hard mental work. In fact, if you don't find practice tiring, you're probably not doing it right.

6. You need to **plan for the long haul.** Complex skills take a long time to master. How much practice you will need depends on which skill you're working at and the efficiency of your practice sessions, but you should be thinking in terms of years, not weeks or months.

You may find this list a bit depressing, given that I've just said becoming good at something is really hard work and takes a really long time. I offer some ideas on motivation in chapter 10, but also keep in mind that there's an amazing prize at the end of that long, hard road. And, of course, there's pleasure in achieving smaller goals along the way.

In a sentence: If the purpose of the activity is to improve your performance of the activity, simply doing it repeatedly isn't enough; improvement requires deliberate practice.

TIP 18

If the Main Point of the Activity Is the Experience, Plan What to Observe

Some activities are irreplaceable for learning. Riding along with police for an overnight shift, visiting an inpatient psychiatric ward, observing a platoon under fire: these are all examples of activities you could read about at length but still would not have much of a feel for until you were part of them. Such experiences can be life-changing.

The depth of the experience gives this sort of learning its appeal, but it can also be the drawback. The environment may be so absorbing that you watch as though you were at a movie. Later, it's hard to say much about what you saw and heard, other than that it was compelling.

You can minimize the chance of that happening by **developing an advance plan of what you hope to learn**. If you accompany an attorney as she confers with clients who are awaiting trial in jail, you might focus on how she talks to them about the future. If you're shadowing a physician, perhaps you want to focus on the way she explains complicated medical ideas to people without training.

If a learning experience is part of your schooling, there's likely an assignment coupled with it. You'll be asked to write about your thoughts or answer specific questions about your experiences. If so, **the assignment should influence what you observe**. Think about the assignment before you go to be sure that you'll be able to complete it, based on your experience.

Often the assignment is vague, something such as "Write a two-page reaction paper describing your experience." In that case, try this: write down your thoughts before you go. What do you expect to see? What do you expect to feel? To what would you compare the place you're to visit? Do you think you'll want to return? What do you think the people will be like? What will they be doing? I still recommend selecting something to focus your observations on—the people, the locale, something—but these predictions will make it easier to shape a reaction paper. You can write about the contrast between what you expected and what you experienced.

In a sentence: If the purpose of the activity is to experience something because you cannot learn about it any other way, plan for what you will observe, because the experience may be so absorbing that you'll otherwise take little away from it.

TIP 19

Don't Forget to Take Notes as You're Experiencing

Unfortunately, the advantage of learning activities—they are interesting!—also makes you less likely to take notes. You may be so engaged in what you're doing that you forget to jot down your thoughts about it. Even if you do remember that you're supposed to take notes, doing so may seem unnecessary because the activity seems unforgettable.

More likely, you'll remember some of what happened and your emotional reaction to it. What you're likely to forget are the insights you had. If you're observing a preschool classroom, for example, you'll remember that a little girl smacked a little boy who accidentally knocked down the block tower she was building, and the double meltdown that followed. You might remember that she was the same little girl who couldn't wait for everyone else to be given their cookies (as she was supposed to do) before eating her own. What you'll forget is your thought that those two issues might be connected and your intention to ask your instructor about it. (They are both examples of difficulty controlling impulses.)

Take notes—during the activity if that's possible or right afterward if not. Remember the memory function of note taking: it will sharpen your focus and force you to verbalize what you're learning. You will likely be tested on the content you were to learn during the activity or be asked to write about it, so you'll need to take notes as a reminder of what you learned.

Tips 14 through 18 emphasized that instructors select learning activities for different purposes and that the purpose of each should guide what you direct your attention to during the activity. Note taking during an activity will probably be hurried, so it's wise to **write the purpose of the activity at the top of the page**. That will help you remember to pay attention to and take notes on that particular aspect of the experience. And if you're worried that you'll forget to take notes, consider setting your phone to vibrate every ten or fifteen minutes as a reminder.

In a sentence: Taking notes during a learning activity may feel awkward and unnecessary, but you should do it anyway, if not during, then right afterward; the laws of forgetting aren't suspended when you learn this way.

TIP 20

Look at Things from the Instructor's Perspective

Assigning learning activities to be performed in class makes many instructors nervous, for a few reasons. First, we feel that we're surrendering control. When an instructor is lecturing, he knows he's *teaching*. He's up there imparting information. But when an instructor gives students something to do, it feels more as though he's hoping they will learn something but can't be sure they will. More than that, an instructor often doesn't even know if students are doing the task he set.

Second, it's hard to come up with good activities. Students must find the activity interesting and challenging but not too difficult, and they must learn something from it. Even activities that were successful in one class may not work with another; the students may differ in their knowledge or interests. In truth, most of the time instructors have no idea what happened. We just know that the 9:30 class killed it and for the 2:30 class it was a dud.

Third, instructors feel anxious during activities because they must juggle a lot of demands on their attention. The instructor tries to monitor everyone's progress, help individuals (or groups), keep her eye on the clock, and judge whether things are going as planned or she needs to supplement the activity with an explanation.

As a learner, you will get more out of the activity if the class goes smoothly, and there are ways you can help ensure that that happens beyond the obvious steps of paying attention and making an honest attempt to do the activity.

First, **be understanding if the instructor seems preoccupied by whether or not you're doing what you're supposed to be doing**. It may seem as though we don't trust you, but actually we're just nervous. We want things to go well, and for many activities it's really hard to tell whether you're engaged or not.

Second, **if the instructor forgets to tell you the purpose of the activity, ask politely**. Obviously, the phrasing matters. "Hey, what's the point of this?" sounds hostile, so maybe try "What should we pay special attention to?"

Third, **let the instructor know whether or not you feel as though you're learning**. If you're worried that it will seem as though you're evaluating the instructor, here's a simple way to provide this feedback. Tell the instructor (1) what you've done (so we know that you're trying) and (2) what you think it means (so we know that you're thinking). Something like "I did *this*, and then I did *that*, and then *this thing here* happened. So, based on all that, it seems like I should conclude *thus*. Does that make sense?" In other words, don't just say, "I don't get it." Let the instructor decide whether or not you are getting it. This feedback is very helpful to us, because if a bunch of people don't understand, we can provide more explicit guidance or abandon the activity altogether and try something else.

In a sentence: Conducting learning activities makes instructors nervous; you can make an activity run more smoothly if you provide feedback about how it's going and how the instructor might help you.

For Instructors

There's a consistent theme in this chapter: tell learners what they are supposed to get out of the activities you set for them, and tell them how to get it. That's the best and easiest way to maximize the value of an activity for your students.

Let's look at what this advice might include. Earlier, I advised students to "be there," and part of that includes doing the reading or whatever other preparatory activities you've set—but, naturally, students don't always complete such assignments. You might consider a quiz (or other assessment) that encourages preparation. I make it easy—just show me that you got the main point—and low stakes so it's not stressful, but if a student never does the readings, the points add up.

When it comes to a scripted activity such as a science lab, it's especially important to talk with students about the learning goal. If they need not worry about the activity "coming out right," what *should* they focus on? They will know they are learning the right thing only if they know the goal *and* if you tell them how they can know that they are "getting it."

In addition, be certain that your materials support the learning goal you describe. If the lab directions are unclear or incomplete, it's natural that students will devote a lot of attention to the method; they are trying to figure out what to do. If you want them to think about big-picture stuff, you need to ensure that they either already know the small-picture stuff cold or that the lab directions provide good guidance.

For projects, be aware that the average American student has little or no experience with them. He will be a rank beginner in things such as selecting a project goal, planning and scheduling, responding flexibly to unexpected problems, and so on. You probably ought to view project planning and execution as a large piece of what you're teaching. Students need to be shown how to break a project down into manageable steps, and they'll need feedback on each step.

The same is true of group projects; most students have little experience working in groups and therefore don't know what it takes to be a good group member. Unsurprisingly, then, students typically worry about other students carrying partial responsibility for their grades and fear that they will end up doing other students' work. A popular meme on social media a few years ago showed a pie chart titled "What Group Projects Teach Me," with a tiny sliver of the circle labeled "The Material," another sliver labeled "Group Skills," and the bulk of the circle taken up by "How Much I Hate Other People."

To calm students' fears about grades, I recommend some formal mechanism of accountability. Here's my approach. At the end of the project, students rate one another (and themselves) on (1) the difficulty of the task undertaken, (2) how hard the person tried, and (3) the quality of the work contributed. I tell students that these assessments may change individual grades, and it seems to make students take their responsibilities more seriously.

Activities are great: students find them engaging, and there are things that are best learned through an activity. But don't fall prey to the idea that because students will be doing something (and you won't be lecturing), this sort of instruction means less work for you. In my experience, setting up an activity, guiding it, and assessing its effectiveness demands much more work than lecturing does.

Summary for Instructors

- Tell students what they are to learn from the activity. Tell them what to pay attention to.

- If the activity requires preparation, consider a preactivity quiz or other assessment of students' preparation.

- If students worry about doing the activity "the right way," and that's not what you want them to focus on, give them explicit instructions for the activity *and* give them a concrete way to know whether the activity is going okay. Or convince them that the outcome doesn't matter.

- If you assign a project, assume that you need to teach students how to manage a project.

- If you assign a group project, assume that you need to teach students how to be good group members. Also assume that you need to assuage fears that some students will shirk.

CHAPTER 4

How to Reorganize Your Notes

A 2007 survey of college students showed that about half agreed with this statement: "My notes are disorganized and hard to understand." In my experience that probably means that about half of students don't realize that their notes are disorganized.

I'm only sort of kidding. Even if you carefully follow the advice I offered in the last two chapters, your notes will probably be just okay, because, as I've emphasized, taking notes is a difficult mental task. You need to revisit your notes to make them more useful.

Remember that in chapter 1 I explained that listeners are likely to notice novel facts and definitions as things they should record in their notes. They also get broad themes because they are repeated, but they often miss the *connections* among facts and ideas.

In chapter 1 I also explained why good organization is so important to understanding, so it's obvious that one reason to revisit your notes is to ensure that you understand everything. But there's another important consequence of good organization: it makes content much easier to remember.

Good Organization Helps Memory

A classic experiment illustrates the importance of good organization to memory. A group of subjects was told that they would be shown twenty-six words and should try to remember them. Half of the participants saw the words logically organized as a tree diagram, similar to the figure below.

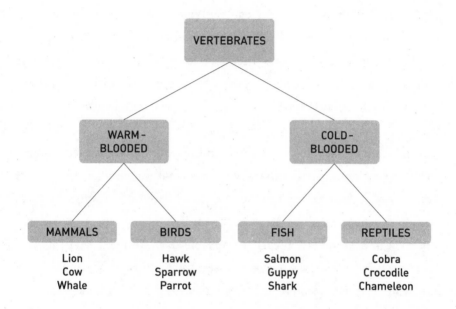

The other half of the subjects saw the identical set of words, also arranged in a tree diagram, but their positions in the diagram were randomized so that the organization made no sense.

Everyone had been told to remember the words and not to worry about their organization. Nevertheless, people seeing the organized version remembered 65 percent of the words, and those seeing the unorganized version remembered just 18 percent. **Organization creates links among the bits and pieces of what you're trying to remember.**

Here's another example. Suppose I asked you to remember this list of words:

Apple Bears Dogs First Leaves Male Next Phone Pilots Smoke

The task would be much easier if you organized them into a sentence, even one that doesn't make much sense.

First, dogs smoke apple leaves; next, pilots phone male bears.

If I can remember "pilots," that's a cue to remember what the pilots were doing: phoning. And if I can remember that they were phoning, that's a cue to remember who they were phoning: bears. And so on.

If reorganizing your notes after class is such a good idea, how come no one does it? In part because evolution has left you with a bias about what is worthy of attention. The brain thinks you should attend to novel things. Familiar things are safe—if they have not been a threat in the past, they are unlikely to pose a threat in the future. Thus, **we see no need to pay attention to the familiar**. We know what the familiar thing is, we know it won't do anything to us, and therefore your brain wants to move on. That's why you quickly grow impatient if you look over the notes that you took at a presentation earlier the same day. "Yes, yes," you think, "I know all this." Your brain tells you to seek new information. But this is a case when learners would benefit from disregarding the brain's impulse to seek out what is novel.

WHEN REORGANIZING YOUR NOTES

What your brain will do: It will conclude that there is no point in reviewing and reorganizing your notes because the content seems familiar.

How to outsmart your brain: Ignore your brain; you know that both the information and the organization are incomplete.

People often see reorganizing their notes as a waste of time; it seems like a prelude to the real work of studying. But **reorganizing your notes doesn't just make it easier to study; the process of reorganizing notes is itself studying.** Reorganizing forces you to manipulate information and think about meaning, and, as you saw in chapter 3, you remember what you think about.

But how exactly are you supposed to reorganize your notes? Let's take a look.

TIP 21

Find Connections Among Elements of Your Notes

In chapter 1 I emphasized the likelihood that you'll miss the deep structure of a lecture when you listen to it; ideas that have some connection (B was caused by A or B is an example of A) may be separated in time, and you may miss their relationship even if you are listening for such connections. In this chapter I've pointed out that appreciating the organization of the material not only is essential to fully understanding it but also aids memory. You want to be sure that you have grasped all aspects of the material's organization. Explicitly **re-creating the full logical structure of the lecture is the best way to do that.**

Typically, this structure forms a hierarchy: one main point of the lecture with three to seven subpoints, each with supporting evidence. For example, you might attend a lecture with the theme "The Myth of the American West Arose in the Late Nineteenth Century" with three main subpoints, each corresponding to a false belief held by American easterners: (1) The West was seen as a foreign country, even though it had extensive communication and commercial ties with the East; (2) the population of the West was thought to be primarily white, even though immigration was very diverse; and (3) easterners thought the West was built by rugged individuals without support from cities, electricity, or industry, when all three actually played important roles in changing the West. Each subpoint has examples, references to conclusions drawn from other lectures, explanations, and so on.

It's a good idea to **draw a tree diagram** rather than just think about the organization. Trying to see it in your mind's eye will be overwhelming—there's too much information, so you'll forget facts and confuse yourself. Commit the idea to paper or draw it on a computer screen using boxes to represent statements such as "The myth of the American West arose in the late nineteenth century." Use lines to connect related statements. For example, the statement "The population of the West was thought to be primarily white" is connected to the statement "The myth of the American West arose in the late nineteenth century" because it's an example of that broader theme.

I encourage you to think about two things as you're trying to figure out the boxes and the lines that should make up this sort of hierarchy. First, **the statements in the boxes should be specific**. For example, the theme of the day is not "The American West"; it's much more specific. People tend to fall back on generalities when they learn something new, partly because, being general, they are safer. For example, if I ask my students to write an outline for a paper, they often make the first entry of the outline "Introduction." In my experience, students think that generalities sound more academic, more learned. Actually, the opposite is true. Your instructor is going to present the content with a particular point of view. You won't get a sequence of disconnected facts but an argument that builds to a conclusion.

Second, when you build the hierarchy, **be specific about why you are**

connecting statements. I suggest you label the lines linking the boxes; typical links would be:

- Provides evidence
- Example
- Elaboration
- Cause
- Logical implication

Reorganizing your notes provides still another advantage. It can be hard to see connections between an instructor's lectures and the readings. As I mentioned in chapter 1, the instructor usually tries to straddle a line between offering completely new content in a lecture (which can confuse students) and restating what's in the readings (which strikes students as pointless and boring). Hence, there is often some overlap, but only some. With the reorganization of your notes completed, you'll have a bird's-eye view of the lecture; that's the time to **consider how the lecture relates to the readings**. (I'll discuss how to get the most from readings—including how to take notes on readings—in chapter 5.)

> *In a sentence:* You've tried to understand how the ideas of a lecture were organized as you took notes, but that effort was probably incomplete; after the lecture, draw a diagram that illustrates how the main ideas of the lecture related to one another.

TIP 22

Spot Holes in Your Notes

After you've built a tree diagram representing the logic of the lecture, you're in a better position to **identify what's missing from your notes**. You're obviously better off trying to recall this missing information sooner rather than later. Ideally, then, you'll do this reorganization of your notes the same day that you took them.

The missing information that you ought to flag falls into one of two categories: **facts** and **connections**. You will already have jotted notes to yourself during the lecture when you missed something the lecturer said or you didn't understand the point of a long explanation. You will not have noticed everything you missed, but it's a start.

Building your tree diagram of the lecture should make other missing information evident. For example, maybe you have a fact recorded in your notes and you have no idea why it was mentioned—it's marooned, disconnected from everything else. There, in a lecture on the book of Elijah in the Bible, is the definition of a *wadi*, along with an explanation of how wadis form. You dutifully wrote it down, but you don't know why the instructor brought it up.

In addition to missing connections, look for missing content. If your notes say, "5 types of map usually used 4 surveying," you'd better have five listed, not four. Look, too, for the number of supporting points for generalizations the instructor made. Suppose your notes say, "Fall of Roman Empire often dated 476 because emperor gone—FALSE. Cult, econ life cont." Okay, so cultural and economic life continued after the emperor was overthrown. But that's just one bit of evidence that dating 476 as the fall is inaccurate. You should be suspicious that your notes list only one source of support for such a broad statement; it is likely that the lecturer offered more.

Write questions about your notes on Post-its and place them so they protrude from the pages. You'll recall that I suggested you write on every other page of your notebook (see tip 10). The purpose is for you to have plenty of blank space to add missing information close to where it belongs in the lecture—information you may have once you answer the queries on your Post-its.

If you took digital notes, you can type your questions directly in your notes, right where you noticed the missing information. To make the questions easy to find later, add the letters "TK" at the end of each question, then you can simply search the file for "TK." ("TK" represents "to come" but is also used because the letter sequence appears in very few words and thus is easy to search for. Pick another letter pair if you take a course on the Atkins diet.)

How do you get answers to these questions? Assigned readings may be one source, but the instructor may mention details that don't appear in the readings. **Be careful with sources not assigned by the instructor**, because there may not be universal agreement on the "facts" you're looking for. The definition of *psoas muscle* will be the same in most sources, but the answers to "Why is 476 the wrong date for the fall of the Roman Empire?" will not.

Your next step should be to consult with other people who took notes on the lecture. If you're concerned about how to reach out to fellow students, keep reading.

In a sentence: Use the lecture organization you've derived to identify what your notes are missing, both facts and connections.

TIP 23

Consider Note Taking to Be a Team Sport

I expect it's obvious how a study group can be useful to improve your notes: you can **compare the organization** that you derived (see tip 21) to the ones that others came up with and decide whether yours can be improved. You can also **fill factual gaps in your notes** that you've identified (see tip 22). Each lecture attendee may miss 50 percent of the content, but each person will miss different things. Here are a few tips on organizing a study group.

Meeting weekly typically makes sense, as one week's worth of lectures is usually a good volume of notes to tackle in a single session. In addition, if you try to meet on an as-needed basis, you'll probably struggle to find a meeting time.

Three to six people is a good group size. You want some variety in perspectives, but with too many participants you might observe what psychologists call "diffusion of responsibility." That's when people don't do what they're supposed to because, the group being large, each person figures that someone else will meet the responsibility.

Identifying potential group members is easy if you know some people in the class. If you don't, you have a few options. You can, of course, simply approach people you meet, or there may be an electronic blackboard for the class on which you can post messages to solicit interest in a study group. You can also ask the instructor to make an announcement that people interested in forming a group should stay for a few minutes after class to organize one.

Identifying people you'll be *happy* to work with is another matter, and of course some people are better group members than others. When it comes to group members not following through (not showing up, failing to prepare, or ignoring group communications), laziness is much worse than incompetence. People will forgive others when they don't contribute much if it seems that they are trying hard. But if they simply seem not to care, there's resentment.

The best way to handle this issue is to **address members' responsibilities at the start**. Set some basic ground rules concerning expectations: how often you will meet, what sort of preparation everyone is supposed to do, if it's okay to use your phone during sessions, who will lead sessions, and so on. If you have this conversation the first time the group meets, one or two people may roll their eyes, thinking, "This seems a little intense." Maybe it is (and certainly, you can be pleasant about these issues), but people can have widely different expectations about how such a group functions, and making expectations clear from the start will make the group's work much smoother.

I've said that missing a lecture but "getting the notes" is no substitute for being there. The notes you're reading are memory cues that someone else created for their own use. But your instinct that someone else's notes are better than nothing is absolutely right. And getting notes from three people is better than getting them from one.

That said, don't make this a set arrangement. In other words, **don't divide assigned work among group members, who then share their efforts**. Think of your team's efforts as an added benefit that will improve your thinking, not as a way to reduce your workload. Creating outlines is not busywork that enables the real work of studying; it *is* studying—it's a cognitive task that will help you understand and remember course content. Getting an outline from someone else does not require mental effort from you and will not yield the same mental benefit.

> *In a sentence:* Join or organize a study group to help fill gaps in your notes and fine-tune their organization.

TIP 24

When Getting Help with Your Notes from the Instructor, Ask Focused Questions

I've suggested that you not only show up to lectures, listen carefully, and work hard to take good notes but also reorganize your notes (the same day if possible) *and* review and further refine them with others. Despite all of this effort, you might still find yourself unclear on some points. That's the time to talk to the instructor. This prospect makes some students nervous, and it's true that instructors vary in how welcoming they are. We're all required to say something like "Come see me with questions anytime!" but some instructors may as well have a sign hanging around their necks saying "Leave me alone." Here's how to placate a grumpy instructor if you need to ask questions.

The key is preparation. Suppose I just spent forty-five minutes lecturing about the retina, and you come to my office and say, "So . . . the retina. I really didn't get that." That's kind of depressing for me, because I don't get the impression that you've done anything to try to understand. **Focused questions** show that you're doing your part to learn. Be prepared to (briefly) **tell the instructor what you did understand** and what pieces are missing for you. Remember I said that if someone in your study group is struggling, you're fine with it so long as you think they are trying? Instructors are the same and probably more so.

Students sometimes bring questions to instructors not because they really need help but because they think it makes a good impression—they're showing interest! Others with the same motive pop in just to chat about topics unrelated to the course or to offer some flattery about how fascinating we make the content.

Students aren't completely wrong; it's useful if the instructor likes you. But **if you're going to suck up, suck up by being a model student**. Be the student carrying a hyperorganized notebook with a dozen Post-it notes peeping out, each bearing a question about the course content. Or come with a set of questions about career opportunities in our field. Even if we suspect that you're playing a part, at least you're playing the right part. We're not going to raise your grade just because we like you, but if something goes wrong and you need a deadline extended, or if you later want a letter of reference, it helps if the main thing we remember about you is that you were earnest about the right things.

In a sentence: If you feel shy about asking the instructor for help in filling your note-taking gaps, come prepared to describe what you *did* understand; good preparation is the surest way to earn an instructor's goodwill.

TIP 25

OPTIONAL: Make Your Notes Look Good

Based on social media posts, the most common way that students review their notes is rewriting or embellishing them to make them more attractive. Search YouTube, Tumblr, or Pinterest, and you'll find thousands of resources (some of them with millions of views) devoted to techniques to make class notes pretty. Gorgeous fonts, page borders, fancy section dividers, boxes you can draw around headers—the energy and creativity people put into the task is impressive. But does it help you learn?

From a cognitive point of view, we'd predict that **the act of copying**

itself does nothing to improve your understanding or memory. In chapter 2 I mentioned that trying to write notes that capture a lecture word for word can lead to shallow understanding; the words go straight from ear to pen, so to speak. The same can be true when you copy your notes; you rewrite the words without thinking about what they mean, whereas it's thinking about meaning that helps memory. My students who like to copy their notes tell me they think it helps their memory, and that's probably because they *are* thinking about the meaning as they copy. But copying is less efficient than the methods of learning we'll examine in chapter 6.

You might couple the copying/beautifying of your notes with the reorganization that I advocated in tip 21. I think the people who enjoy this process would argue that having attractive notes makes them more eager (or at least more willing) to study later. Again, I know of no research on the matter. So if you feel that drawing borders and boxes and using multicolored headings is helping you, I'm not going to deny it. Just be aware that simply rewriting won't help your memory or understanding; it's other cognitive processes that you're engaging while you copy that are providing the cognitive benefit, and there are more straightforward, reliable ways of engaging those cognitive processes.

In a sentence: Beautifying your notes won't help with your comprehension or memory, but if you enjoy the process or the result, there's certainly no harm in doing so.

For Instructors

You know that your students' notes aren't perfect, but you also know that they will be reluctant to review them as I suggest. How can you help?

One obvious measure is to alert students to the problem. Most don't realize how spotty their notes are, so spend fifteen minutes of one class meeting outlining the problem. You might start with a low-stakes pop quiz during which students are allowed to access their notes and where all the questions come from a single lecture. Once students see that their notes are incomplete, describe the methods in this chapter by which they can address the issue. You could also consider having frequent open-notes quizzes to motivate students to keep their notes current and complete.

You can facilitate the creation of study groups by announcing during class that interested students might stay a few minutes after the lecture for this purpose. That's easy for you, but leaves out students who can't stay or who missed class that day. Emailing everyone is a more reliable method; you can either collect the names of those interested and then email the list to them or use a website or app (for example, GroupMe) that serves this purpose.

But if you know your students' notes are incomplete, shouldn't you do something to ensure that they can learn all of the content? I don't think there is a single right answer to this question; each option has advantages and disadvantages, as listed in the table below.

Possible Solution	Advantage	Disadvantage
Do nothing.	Is easy for you. In truth, most schools don't expect you to support note taking, because most people don't realize it's a problem.	Some students don't learn as much as they might. Gives an edge to students who write fast and to those who start the class already familiar with some of the content.

Possible Solution	Advantage	Disadvantage
If recitation/review sessions exist, use them for clarification and the answering of questions.	Is easy for you. You get feedback from recitation section leaders about students' confusion.	Recitation sessions could be put to other uses, e.g., discussion or application of ideas.
Count on a textbook or other readings to fill gaps.	Students feel confident that they have a backup in written materials.	Because lectures and reading overlap, motivated students may be bored in class, and unmotivated students may skip.
Answer questions about content via an online forum.	Provides feedback to you about students' understanding. Ensures that students get the content right.	Is time-consuming for you.

If students' notes are so incomplete, perhaps we shouldn't plan for them to learn in classroom situations where they get just one shot at understanding. Maybe the bulk of learning should happen via reading or video, which students can review as often as they wish. Wouldn't that greatly reduce or eliminate holes in students' notes (and their understanding)?

A lot of college instructors got a crash introduction to that method during the COVID-19 pandemic. We had to record lectures because many of our students were in different time zones, so lecturing live was impractical. A few students thrived under the arrangement, but most had a lot of trouble motivating themselves to watch the videos.

Naturally, this was not a fair test of the method, because the students were under enormous stress during the pandemic. But it wasn't the first time recorded lectures had been tried. There was a big push for this method around 2010. At that time, the argument was that if students learned the basic content at home, via video, they could work problems or have a discussion in class, when the professor was there to help. But then, as during the pandemic of 2020–2021, a lot of students didn't watch the videos.

Students seem to want a live person to explain content despite the note-taking challenge that comes with a lecture. We must address the inadequacy of the notes students take as best we can.

Summary for Instructors

- Use the tips from chapters 1 and 2 to ensure that students get as many facts and as much of the organization as possible.
- Let students know that their notes are probably incomplete and disorganized; they won't be motivated to improve them if they don't perceive the problem. Consider low-stakes, open-notes quizzes for this purpose.
- Facilitate the formation of student study groups.
- Consider carefully how, if at all, you will supplement students' notes after lectures; there are advantages and disadvantages to each choice.

CHAPTER 5

How to Read Difficult Books

It seems obvious why textbooks are hard to read. The material is dense; there's a lot of information packed into relatively few words. Authors often feel obligated to give you a broad, complete understanding of a topic rather than to weave an interesting story from selected details. Teachers are ready to assign a textbook, even if it's boring; it's seen as a regrettable but unavoidable problem.

But there's a more subtle reason that it's hard to stay engaged when you read a textbook. To find out why, read this paragraph, one you might find in a typical high school textbook.

The Manhattan Project was the United States' effort to produce a nuclear weapon, and it was the largest construction enterprise in the history of science. Because of its sensitive nature, a massive effort was made to keep the project secret. Famous scientists traveled under aliases; Enrico Fermi was known as Henry Farmer, for instance. And all telephone conversations at the test sites were monitored. Despite those efforts, historians agree that it probably would have been impossible to keep the secret if not for the fact that the project was of relatively small size.

Did you notice that the last sentence contradicted the first? Embedding a mistake or contradiction into a text and seeing whether readers notice it is a common research technique to measure comprehension. Readers are asked to judge each text on how well it's written and explain their rating.

Readers are *very* likely to notice a word they don't know. They are also very likely to notice if the grammar of a sentence is wrong. But they are **much less likely to notice when two sentences contradict each other**. Forty percent of high school students missed the contradiction in the paragraph above. To put it another way, if readers simply understand each sentence on its own, they figure they are doing what they're supposed to do.

Coordinating meaning across sentences is crucial to reading comprehension, because sentences can take on quite different meanings depending on the surrounding context. For example, consider a simple sentence, "Maxim waved," in different contexts:

Ann walked into the pizza parlor, looking for her friends. Maxim waved.

The boat slowly circled the wreckage, looking for survivors. Maxim waved.

"Oh, my God, that's my husband!" Kate whispered. "Don't do anything that would attract his attention!" Maxim waved.

In one way the sentence always means the same thing—the physical act of waving by Maxim—but the more important meaning—*why* Maxim waved and the likely consequences of his action—is very different. It can be appreciated only if you interpret the sentence in light of what you've already read.

In chapter 1 I said that lectures are hard to understand because they tend to be organized hierarchically, so related ideas may be separated in time. I also said that you can't benefit from a lecture if you just sit down as though you're a member of an audience about to watch a movie, expecting an entertaining story. The problem is that we're biased to listen exactly that way.

The same problem applies to reading textbooks. Writers organize the material hierarchically, so **readers often need to connect what they're reading now to something they read a few pages ago**. But readers, like listeners, expect a simple format. We first learn to read storybooks. Stories are easy to understand because the structure is simple and linear: A causes B, which causes C, and so on. Textbooks are more like lectures in their hierarchical format and challenging content. Yet just as we tend to sit down in a lecture and expect to be entertained, so, too, we sit down to read a textbook and expect that the author will make our job easy. You need a different approach to reading such content.

WHEN READING TO LEARN

What your brain will do: It will read the way you read for pleasure, because that's familiar to you and it's not obvious that it won't work. You'll read making minimal effort to coordinate ideas, trusting that the writer will make the connections explicit and easy to follow.

How to outsmart your brain: Use specialized strategies for comprehension that fit both the kind of material you're reading and the goals you have for reading it.

Learning by reading is a substantial challenge, but with a few strategies under your belt, you can be much more successful in connecting the ideas as the author hoped you will.

TIP 26

Don't Do What Most People Do: Just Read and Highlight

Let's start with the most common tactic people use when reading with the intention of learning. They open the book and start reading. When something strikes them as important, they mark it with a neon highlighter. They believe that the act of highlighting will help affix the information in their memory and that highlighting creates a ready-to-use study guide. Later, they believe, they can refresh their memory by rereading what they've highlighted.

This is a terrible plan. It does not address people's habit of failing to coordinate meaning across sentences and paragraphs. How can you be sure that you are highlighting the most important information if your understanding is hit-or-miss in the first place? Furthermore, even if you understand everything quite well, how can you be sure that you are a good judge of what is important enough to highlight as you read the content of a topic you know little about *for the first time*?

Both problems—you may not understand as well as you think and you may judge importance poorly—suggest that **people don't highlight the most important information**. Researchers have tested that prediction with a simple, clever method. They went to a college bookstore and bought ten used copies of the textbook for each of three courses. If spotting the most important content were easy, everyone should have highlighted the same material. But the researchers found little overlap in what students had highlighted. That's why I've used boldface type for the important points in this book; I've done the highlighting for you.

Please note that this advice doesn't mean "never highlight." **Highlighting might be fine if you are reading about a topic you already know a lot about.** If you've been a political consultant for twenty years and you're reading a briefing on a recently concluded statewide campaign, your deep knowledge of the topic means that you will read the document with good comprehension and your knowledge will also make you a good judge of what information in the document is important.

A college student reading the same document as part of a political science course lacks the necessary background knowledge, but there's another reason the political consultant reads the document with better comprehension: she knows what to expect. She knows the type of information such a document usually contains, and she knows the function it's meant to serve. The novice doesn't.

If you have even a vague idea of what to expect when you read, that will make you read differently. You'll notice and remember different details, for example. A chapter on the Human Genome Project, the effort to map all the genes in human DNA, might focus on any of several aspects of such a complex topic. It might describe the expected economic benefits to the pharmaceutical industry or the project's impact on gene therapy. It might describe the politics of the government's funding such a huge project. Knowing the author's goal before you begin reading gives you a start on evaluating which ideas in the chapter are most important.

Thus, highlighting is not the only flaw in the "just read and highlight" approach. "Just read" is also a bad strategy, because **you shouldn't plunge into a text without some preparation.**

Now let's consider what you *should* do.

> *In a sentence:* Reading and highlighting is a poor strategy because it fails to provide a framework for understanding before you read and it leads you to decide that some material is more important than other material, even though you have little basis for that judgment.

TIP 27

Use a Reading Strategy That Fits Your Goal

Tip 26 emphasized that you can't just start reading; that's like attending a lecture as if it were a movie. You need to bring something to the process rather than just wait for the author to intrigue you. At the same time, the advice "Read actively" is nearly useless. You may earnestly set the goal "I'm really going to think as I read, and I'm going to connect ideas," but it's just too easy for your attention to drift.

The solution is to **set a concrete task to be completed as you read**. The best known is called SQ3R, which has been around in various versions since the 1940s. SQ3R is an acronym for these steps:

Survey: Skim the reading, looking at the headings, subheadings, and figures. Get a rough idea of what it's about. This is how you'll determine, for example, that an article about the Human Genome Project is about its economic consequences, not the ethical implications of sequencing human DNA.

Question: Before you read, pose questions that you expect the reading to answer. Headings can be especially useful for this task; for example, if you see the heading "Marr's Contribution to the

Philosophy of Science," the obvious question to ask is "What was Marr's contribution to the philosophy of science?"

Read: Keeping in mind the rough idea of the article's content you developed when you surveyed the reading, it's time to actually read. And now you have a concrete task to be completed as you read: look for information that answers the questions you've posed.

Recite: When you've finished each section, recite what you've learned as if you were describing it to someone else. Summarize it and decide if it answers any of your questions.

Review: Reviewing is meant to be an ongoing process in which you revisit the content, focusing especially on the questions posed and the answers you derived.

Research confirms that using SQ3R improves comprehension, and it's easy to see why. I've explained why you shouldn't just plunge into a reading; if you first consider what it's about and why you're reading it, you will actually read it differently. The Survey and Question parts of SQ3R get you to do exactly that. I also emphasized that it's essential to build meaning across sentences, and reading with the questions in mind also helps accomplish that.

The Recite step of SQ3R ought to help you pull your thoughts together and retain content, but even more, it's a check of your comprehension. Remember that people can easily fool themselves into thinking they understand when they don't. Reciting will help you better evaluate your comprehension.

The one drawback to the SQ3R method is that you may slip into "just reading" without thinking much. Here's a trick that might help: after you've posed your questions (and before you start reading), **place some blank Post-it notes in the text**—maybe one at the end of each section. They'll serve as visual reminders that you should stop, try to summarize the section you've just read, and think about whether the section answered any of the questions you posed.

SQ3R is useful and it's the best known of this sort of strategy, but there are others, including KWL (think about what you Know; what you Want to know; what you've Learned), SOAR (Set goals; Organize; Ask questions; Record your progress), and others. It's no accident that most reading strategies have two important properties in common: they get you to **think about your goal for reading before you start** and **connect the pieces of the reading** by asking big-picture questions.

If these strategies seem like overkill, let me offer an alternative with just one step that may be an easy start to this kind of work. Instead of posing questions in advance, **pose and try to answer questions as you're reading**, especially "Why?" questions in response to stated facts. For example, when you read, "The president can propose legislation, but a member of Congress must introduce it if it's to become a bill," you might ask, "Why must a member of Congress introduce it?" "Why?" questions tend to lead you to deeper principles and connections, in this case perhaps to the idea of the balance of powers among the three branches of the US government.

The advantage of this method is its flexibility—you don't commit yourself to a set of questions before you've started reading. In addition, it's easy to adapt this strategy to readings that tell you how to do something rather than telling you a bunch of facts. How-to information tends to occur in stages, so you can ask, "Why does this step come next?" The disadvantage of this method is that you can't pose a question to yourself every time the author states a fact—that would slow you down too much—so effective question posing takes some practice.

Again, there's no definitive evidence that one strategy is superior to another. What the evidence shows is that **using a strategy is better than not using one**.

> *In a sentence:* Good reading strategies prompt you to think about the content and set concrete goals for what you're to learn before you read, and to connect ideas as you read.

TIP 28

Take Notes as You Read

Whenever I meet with a student who is struggling in one of my classes, I always ask her to bring her notes. Everyone has notes they've taken in lectures, but most people do not take notes on the readings. Surveys bear out my experience. People don't take notes on readings because they figure that highlighting serves the same purpose. But we've been over why it doesn't.

Taking notes on readings serves the same functions as taking notes during a lecture: **it helps keep you mentally on task, and the notes will help refresh your memory later**.

But there are differences in the note-taking process when you're reading. The most important is that you, not the speaker, control the pace. You can read as quickly or slowly as you want, and you can revisit older content or peek at what's to come later. This removes one of the main concerns about using a laptop to take notes while reading. During a lecture there is the risk that the urge to keep pace with the speaker will prompt you to go into dictation mode. Since that problem is irrelevant when reading, I would be more likely to **take notes on a laptop** because they are so much easier to edit than handwritten notes and later to search for information. Naturally, you may have other reasons to prefer paper: you find social media hard to resist, for example, or you need to draw a lot of figures in your notes. Or you simply like paper better. It's up to you.

How should you begin? In particular, how should you prep for taking notes? The same way you prep for reading, by posing questions at the start. But how can you craft good questions about a text you haven't read? The author may give you a good overview in the first few paragraphs, or perhaps there are questions at the end of the reading that provide some guidance. Or maybe the instructor, God bless her, told you what she hoped you'd get out of the reading. Write these at the very top of your notes, so you can keep them in mind as you read.

If the reading includes **headings and subheadings,** you might write those in your notes; they can serve as a skeletal outline. As you read, complete the outline. **For each subheading, write a summary and about three other statements.** These statements might include, for example:

- An important qualification to the summary
- A comment on how this section relates to the main section
- How the section answers one of the questions you posed for the reading as a whole
- An implication of the summary for something else the author concluded

You should also include any new vocabulary terms and their definitions. As much as possible, use your own words, not the author's; as was true of lectures, there's no point in taking dictation. You need to manipulate the material mentally.

As you consider exactly what to record in your notes, you might **think ahead to how you will use them.** If you'll later be tested, consider that there are different types of test questions. I'll have more to say about this in chapter 6, but for the moment, consider the difference between short-answer and essay questions. Each emphasizes different types of content. Answers to the former are necessarily short and often call for definitions, dates, or examples to be categorized. Essay exams, of course, pose broad questions, so you had better understand themes and how things connect. If you know how you will be tested, pay special attention to the content that's vital for that type of assessment.

When you've finished reading and taking notes, you may be delighted to be through with the job. Actually, you're not quite done. **Once you've completed the reading, you should look over your notes to be sure you're satisfied.** Did you answer the questions you posed? Are you still convinced that they were the right questions? Do you think your notes are good enough that even if you set them aside for a few weeks, rereading them will enable you to recover all of your insights into the content?

Finally—and you don't have to do it right now—if there's a lecture associated with the reading, you should consider how the two relate. If you are virtuous and completed the assigned reading before the lecture, you can try to anticipate. If the lecture has passed, don't let this task be forgotten.

In a sentence: Take notes on the thoughts generated by your reading strategy; doing so will help ensure that you don't mentally drift into casual reading, and the notes will, of course, be useful for reviewing later.

TIP 29

Allocate Significant Time to Reading

It's difficult to read texts on complex topics written by authors who are not afraid to bore their audience. What's more, you're taking multiple classes, and you also have work around the house (and possibly a job) to do. So if reading makes you feel overwhelmed, you should know that you're not alone.

Most school-related tasks—giving a presentation, for example, or taking an exam—carry immediate, obvious consequences if you fail to prepare. But the cost of failing to read something you ought to is usually delayed, so that's the task that is postponed or abandoned.

Some study guides suggest that that's a good idea, and they offer methods to figure out which readings to neglect as well as tactics for skimming those you do take on. Let's start by debunking a couple of common tricks meant to allow you to skip readings.

First, **speed reading is not a thing**. You can waggle your hand from the top to the bottom of the page, but you literally cannot read that fast. Lots and lots of studies have been conducted over the decades showing that people who claim to be speed reading are skimming, and as you'd expect, if you skim difficult, unfamiliar material, you won't understand it very well.

Second, if the readings include learning aids such as chapter outlines, chapter previews and summaries, boldface or italicized terms, or practice test questions, **don't try to use these learning aids as a replacement for reading the text**. The funny thing about these features is that there's very good research evidence that they work. Publishing companies paid to have high-quality research conducted; researchers had people read textbook chapters (with or without the learning aids), and they found that people who used the learning aids understood and remembered more than those who did not.

But the psychologists Regan Gurung and David Daniel pointed out that students "in the wild" will not necessarily use such materials the same way they were used by students in the laboratory. Gurung and Daniel suggested that some students use learning aids not to supplement the reading but to avoid it. They read the summary, look at the boldface terms, and then try to answer the practice test questions to see whether they understand enough to skip the reading.

Now, everyone has times when their schedule backs up or something unexpected happens. I can understand doing selective skimming of a reading when your planning fails you. But as I indicated at the start of this chapter, *planning* to skip readings strikes me as foolish. I've seen study skills books in which the author encourages the reader to adopt this strategy for "secondary" readings. Guessing which readings will be the important ones is like trying to second-guess the stock market; it's not very likely to pay off.

I suggest that you allocate "significant" time to reading. What does that mean in practical terms? In college, you'll often hear "three hours of preparation for each hour in class." A typical college course load calls for 12.5 hours of class time per week, so that rule of thumb means around another 37 hours of preparation outside class (which breaks down to 5.5 hours a day), totaling about 50 hours of work per week total. So a lot, but nothing outrageous. That said, people vary in how quickly they read, and obviously some readings take longer to get through than others.

Although even a rough figure of how much time you'll need is difficult to pin down, you should recognize that reading is the chief way that you will learn in college and beyond. It is worth reading carefully, both to learn now and to develop the knowledge, skills, and habits that will make you a successful reader in the future.

> *In a sentence:* Just as I encouraged you to recognize that listening to a lecture is hard work, so, too, is reading; be sure you schedule enough time to give it the attention and mental effort it requires.

For Instructors

Instructors can help students learn to absorb more from their reading. The techniques you use can follow those I've outlined for students.

First, even middling readers don't see a need to improve. So you might consider a demonstration like the one at the start of the chapter that used the Manhattan Project passage. Clip six paragraphs from materials that you're not assigning but that match the subject matter of the class. For two of the six paragraphs, rewrite a sentence so that it contradicts an earlier statement. For each paragraph students should provide separate ratings for how well written and how easy to comprehend they find it. Collect their responses and see if they have spotted the contradictions.

Second, students will benefit from your modeling the reading strategy. Devote some class time to demonstrating how you would implement it for one of the assigned readings. Even better, stretch this exercise to several readings where you initially provide very explicit instruction in strategy implementation and then offer less support while providing feedback on their attempts.

Even if your students are adept readers, you should tell them the goal for each reading you assign. What do you expect they will learn from it? How does it relate to other readings or topics in the class?

Once your students understand what it takes to read a text deeply, be sure that other policies in your class align with the expectation that they will do that work. If you demand deep reading, you should respect the fact that it's time-consuming. It's fair to trade breadth for depth, so assign fewer pages.

The message that you expect deep reading should also be reinforced by your expectations in class and on assessments. If you say you want students to read deeply but classroom discussions skim the surface, students will quickly perceive what you *really* expect. In my experience students love class discussions that go deep—they are so accustomed to courses that require only that they absorb information that they are excited to feel they understand something in greater depth. Admittedly, they are a little less enthusiastic about assessments that probe deep understanding, but that's another way that instructors communicate the importance of getting beyond assembling facts in memory: they give tests that require analysis.

Summary for Instructors

- No matter how experienced your students are, don't assume they know how to comprehend difficult texts; you may need to teach reading strategies.
- If your students are overconfident about their ability, consider a classroom demonstration to show them that they understand less than they think they do.
- Teach students the strategies described in this chapter, but assume that they will need you to model the process.
- Be explicit about why you assign each reading and what students are to get from it.
- If you want students to read deeply, be sure that the rest of the course aligns with that expectation. For example, the number of pages assigned should be reasonable, and assessments should probe for deep reading, not factoids.

CHAPTER 6

How to Study for Exams

This chapter seems to be the first in the book that's really about *learning*—that is, getting stuff into memory. But although it may have seemed that so far we've talked about *preparation* for learning more than learning itself, you've already encountered two powerful principles of memory: **memory is the residue of thought** and **organization helps memory**.

Those ideas reappear in this chapter, but we'll make the heaviest use of a third principle: **probing memory improves memory**. If you want to cement something into your memory for the long term, it's actually better to test yourself than to study. Let's consider a typical experiment that illustrates the phenomenon.

Experimenters give one group of students a textbook chapter to read and study for an hour. The subjects come back two days later, they are given the same chapter, and again they read and study. Two days after that, they return to the lab and take a test on the content.

For a second group of students, the first and third sessions are identical, but during the second session they take a test on the content instead of studying. The test uses different questions from the ones on the final test, but it covers the same concepts.

The people who take the test during the second session do better on the final test than those who study during the second session, by around 10 to 15 percent.

This is called **retrieval practice**. *Retrieval* is a term researchers use for the process of pulling something out of memory, and the learning benefit comes, obviously, from practice in retrieving something. Retrieval practice works for all ages and all subjects, but there are two limitations you should know about.

First, **feedback matters**. If you take a test for the purpose of learning, you should find out immediately whether or not you got each item correct. If you can't remember or answer incorrectly, you should get the right answer zapped into your memory right away. Second, **retrieval practice works only for what's tested**. In other words, if you read an article about Peter the Great that contains, say, thirty facts about him and then take a test on ten of those facts, your memory has improved for those ten facts but not for the other twenty.

Retrieval practice is also a good example of an effective study method that feels as though it's not working. In the introduction, I offered an analogy to exercise: if you want to be able to do a lot of push-ups, it obviously helps to practice push-ups, but it's better to do really difficult push-ups, such as the ones where you propel yourself off the floor and clap your hands while in the air. Naturally, you won't be able to do as many of these push-ups; you must keep in mind that it's the best practice in the long run even though it feels difficult and you observe that you're not having much success. Your brain will tell you to pick exercises that feel easier and that you can accomplish more readily. That's the challenge of using retrieval practice to commit things to memory: it's hard, and you fail a lot. But it's the right exercise to get things to really stick with you.

WHEN COMMITTING THINGS TO MEMORY
What your brain will do: It will seek memorization techniques that feel easy and that seem to lead to success.
How to outsmart your brain: Use techniques that yield long-lasting memory—organizing, thinking about meaning, and retrieval practice—even though they feel difficult and seem less productive in the short run.

In this chapter we'll look at specific tasks you can set for yourself that exploit the three principles of learning I've described. We'll start by looking at commonly used study strategies that are ineffective.

TIP 30

Avoid These Commonly Used Strategies

Have a look at this list of memorization strategies. How many do you use?

- Repeating information to yourself
- Reading over your notes
- Rereading the textbook
- Copying your notes
- Highlighting your notes
- Creating examples of concepts

- Summarizing
- Using flash cards
- Outlining
- Taking a practice test

Surveys of college students show that these are the most commonly used study strategies. We can evaluate them in light of the three powerful principles of memory we've discussed:

1. Memory is the residue of thought, so thinking about meaning will help.
2. Organization helps memory.
3. Retrieval practice cements information in memory.

Some of the strategies on the list—summarizing, outlining, creating examples of concepts—look pretty good in terms of getting you to think about meaning. Others—for example, reading over your notes, rereading the textbook, and highlighting your notes—don't guarantee that you'll think about meaning. When it comes to organization, summarizing and outlining look promising, but most of the others don't. How about retrieval practice? Using flash cards definitely capitalizes on that principle. Taking a practice test is on the list, but it turns out that people don't use practice testing as a way of studying but as a way of figuring out whether they can stop studying. (And they don't use practice testing the right way for this purpose, as we'll see in chapter 7.)

So some of these strategies are good ones, but unfortunately, **the least useful strategies—reading over your notes and rereading the textbook— are the most commonly used.**

Now, there's nothing to say that you *can't* read over your notes with deep concentration, thinking about the content and making connections as you go. It's just hard to do. Indeed, experiments show that rereading tends not to help memory much. The psychologists Aimee Callender and Mark McDaniel asked college students to read two-thousand-word sections of textbook chapters or articles from the magazine *Scientific American*. People were told that their understanding and memory would be tested later with a quiz or by writing a summary. Some people read the text once, and some read it twice. For the most part, rereading did not help. But rereading is—you guessed it—easy. So you can understand why learners drift toward this strategy.

In a sentence: The most commonly used strategies are ineffective for memorization.

TIP 31

Keep in Mind That Preparing to Study *Is* Studying

I've urged you not to use some study strategies because they aren't a good use of your time; they don't align with the principles of memory we've discussed. The best way to commit information to memory is to think about what it means and make meaning-based connections among all the bits of what you are to learn. Thus, you might suppose that when listing the study strategies you *should* use, I'm going to say that you should set yourself tasks such as outlining and summarizing.

But I'm not going to say that, because **by the time you try to commit things to memory, you should have *already* thought about meaning and organized the material.**

That's the kind of thinking I suggested you do in chapters 1 through

5. I presented the tips in those chapters as helping you to understand new content, and they surely will. To understand ideas, you have to understand how they are organized. And to understand how content is organized, you have to think about what it means.

But let me remind you of one other principle of memory, this one from the introduction: **whether or not you *want* to learn is irrelevant.** All that matters to memory is the mental work that you do, not whether you hope to learn from that mental work. If you follow the tips in chapters 1 through 5, it doesn't matter whether you do so intending to learn; they will still prompt the type of mental activity that is terrific for learning, so learn you will.

The students in my classes who struggle usually don't do the things described in chapters 1 through 5. They also don't really see the point in doing all that stuff. They don't understand that these activities are not simply *preliminaries* to the real work of memorizing; they are *part of* that work.

My students who don't do very well think that "keeping up" in a course means attending lectures and completing the reading on time. It's not until they prepare for an exam that they really think about what all the content means, try to organize it, and try to fill gaps in their understanding. That is dangerously late to undertake that work. Worse, some of my struggling students don't even work on understanding at that point; they just start trying to memorize.

The mental activities that help you understand are study activities, too, and so by the time an exam rolls around, the information you need to learn will likely already be in your memory. You'll still need to study, but you'll have a head start. And the fact that you have *something* in memory means you can capitalize on one of the most powerful study methods: retrieval practice.

> *In a sentence:* The tips in chapters 1 through 5 are designed to help you thoroughly understand what you hope to learn, but in so doing they also provide an excellent start for getting content into your memory and should not be considered optional.

TIP 32

Prepare a Study Guide

I recommend that you write a study guide in question-and-answer format so you can capitalize on retrieval practice. Put into more familiar terms, it will be a massive deck of flash cards. Massive, because the purpose is not only to give you an effective way to study, it's to ensure that everything you need to know is collected in one place. If you're systematic about building the study guide, no exam questions will surprise you. There are three steps to creating and using this sort of study guide.

Step 1: Prepare. Be sure you are clear about the nature of the test. Ask yourself:

- Which lectures will be covered? Which readings?
- Has any information been provided about the percentage of questions coming from lecture versus reading?
- What format of questions will be posed (e.g., short answer, multiple choice, essay)?
- How many questions will there be? Does it sound as though time will pose a problem for you?
- Will you have access to any information during the test (for example, formulas or constants during science tests)?

- Are any aids allowable? A calculator? A sheet of paper to write on? Are you allowed to ask anyone to clarify questions during the test?

If previous tests are available, look over the questions. Unless you're sure that the test is the same every year, don't be too concerned about the content. Pay attention to the **types of questions** posed. Are they straightforward requests for definitions or demands that you apply what you've learned in new contexts? Do they test your understanding of broad themes or your ability to memorize every last seemingly insignificant detail? Are they phrased in a straightforward way, or do they seem unnecessarily tricky? Every test has a range of different question types, but a scan of prior exams may give you a sense of what's typical and what's considered fair game.

All of this preparation should be done **with your study group**. That way you'll be confident that you aren't confused about the information the instructor provided (such as what will be covered), and you'll have multiple judgments about the subjective stuff (such as what old exams are like).

Step 2: Write the study guide. You can use index cards (the traditional flash card medium) if you like or a pad of paper, posing questions on the left-hand side of the page and answers on the right. Or use a digital platform specifically designed for writing flash cards. Studies have compared digital and paper flash cards, and there's no definitive evidence favoring one or the other.

Go through the revised version of your class notes and your notes on the reading, and write questions about all of the content. **Plan to learn everything in the flash card deck but nothing else.** That's how complete you want this resource to be.

Your focus on the *levels of organization* in lectures and readings will pay off again as you write your questions. **Pose questions at multiple lev-**

els of organization and between levels: the lowest level of the hierarchy ("When was the Battle of Saratoga fought?"), the midlevel ("What was the role of the Battle of Saratoga in France's support of the colonies in the War of Independence?"), and the highest level ("Why did France support the independence of the colonies?"). The proportion of questions at each level should vary depending on the type of test—more low-level for multiple choice, more high-level for essay.

Can you really write a flash card for a high-level idea? Sure. Even if the exact question doesn't appear on the test, at least you are thinking about broad themes in the content. Obviously, you are not going to write an essay on the back of your flash card; just write a skeletal outline of what your answer would be. Even if you're preparing for an essay exam, you should still have some low-level questions about definitions, dates, and the like. You'll want to include those in your essays, and it's easier to memorize such factoids if you devote a separate question to each.

It's a good idea to **pose questions in both directions**—that is, to have one question asking for the definition of a term—e.g., "What is opportunity cost?"—and another that asks for the term, given the definition: "What's it called when choosing something means you lose a potential gain from the other alternatives that you didn't choose?" You would think that memorizing a question in one direction would mean that you automatically know the answer when the question is posed the other way, but memory doesn't always work that way. If I ask, "What word comes to mind first if I say *pepper*?" you might say "Salt," but other answers are also common, including "Hot" and "Chili." But if I ask, "What goes with the word *salt*?" you're very likely to say, "Pepper." If you study only one direction and the question is asked the other way, *that's* how you miss the question and think, "How is that possible? I *knew* that!"

In technical courses, generate examples of the types of problems you're expected to know how to solve. You should also have some explanation questions, e.g., "Why is potential energy and not kinetic energy important in this problem?" Maybe you want to add some questions that extend what you've learned, for example, applying concepts to new or real-world conditions. (Your review of previous exams can help you figure out the usefulness of including such questions in your study guide.)

If the exam will include only short-answer or multiple-choice questions, you should be pretty focused on memorizing facts. Your study guide should still include questions that prompt you to **meaningfully connect facts** to one another, not only because you might be tested on them but also because thinking about the connections will make everything more meaningful, and more meaningful content is easier to remember.

Step 3: Commit answers to memory. How long will it take to memorize everything in your study guide? It depends on how much information it contains, obviously, and different people find memorizing more or less difficult. Writing the study guide is a whole lot better than not writing it, and finishing it two days before the exam is better than finishing it the day before the exam. You can continue from there—every day earlier that it's done is a day you can spend a little time reviewing. The main problem, then, is one of *planning*, a topic so important that I devote all of chapter 10 to it.

We'll discuss the nuts and bolts of getting stuff into the memory vault shortly, but first I want to warn against a tempting shortcut.

In a sentence: Make your study guide as complete as possible so there won't be any surprises on the exam.

TIP 33

Avoid Found Materials

Going through all of your notes on lectures and readings and creating questions for all that content sounds like a lot of work. It is. Online vendors sell outlines and flash card decks for textbook chapters and specific college courses; you might also simply get these things from a friend who took the course. You can buy practice tests and other test-prep materials for standardized tests such as licensing exams. Collectively, I call them "found materials": content offered to you as relevant but something that was created not by you, nor by whoever wrote the exam.

I strongly recommend you steer clear of found materials.

For one thing, **found materials often aren't very good**; they contain errors and omissions. Even the materials that come from textbook publishers should be viewed cautiously. They are seldom written by the textbook author, and the instructor may not have thought much about the supplementary materials when she chose the textbook; if you are thinking of using them, ask the instructor whether they will be useful to you. And whether found materials were written by a professional or a fellow student, they may no longer be applicable. I'm constantly updating my classes, and for that reason a deck of questions that was perfect last year won't be perfect this year.

Most important, remember that **writing a study guide is an excellent way to commit content to memory.** That's why I told you not to split the job of creating the study guide among the people in your study group. And that's why I don't want you to use one created by a stranger.

Next up: How can you commit the content in your guide to memory?

> *In a sentence:* Don't use study materials created by someone else; they're often inaccurate or incomplete, and creating your own is an excellent way to study.

TIP 34

Pose and Answer Meaningful Questions to Get Memories to Stick

Okay, you have your sizable, maybe somewhat frightening flash card deck. What's the best way to learn the answers to all of the questions you've posed?

Before I talk about strategies, let me cut short one destructive thought you might have. Don't tell yourself (or me), "I have a terrible memory." Almost everyone feels they have a bad memory, because we notice when memory lets us down. Unless you've been diagnosed with a memory problem by a doctor, your memory is just fine. Yes, I know you have a friend who seems to remember everything with no effort—everybody has a friend like that. Don't compare your memory to the memory of that friend. Yours is good enough; it's a matter of putting it to work.

It's much easier to remember meaningful content than meaningless content. Movie plots are so easily recalled because each scene is connected to other scenes—thinking about Buzz Lightyear falling out the window reminds you that Buzz and Woody end up stranded on the road, which reminds you that they catch a ride on the Pizza Planet truck, and so on. A random list of digits is hard to remember because the numbers aren't connected.

To capitalize on this property of memory, make the answer to each question meaningful, even if the question itself is not a "meaning" question. For example, a question in your study guide may be "What were the years of the Era of Good Feeling in the United States?" If you're having trouble remembering the answer (1817–1825), **make it a meaning-based question by asking "Why?" or "How?"** *Why* does it make sense that the Era of Good Feeling occurred at that time? The year 1817 was shortly after the end of the War of 1812, and there was a strong feeling of nationalism because Americans thought they had won the war. Also, those years coincided with the presidency of James Monroe, who emphasized unity by appointing people from across the political spectrum to government posts.

Asking a "Why?" or "How?" question can make what seems to be an arbitrary link between a question and answer into a meaningful link *and* it will connect what you're trying to learn to other information that you are trying to master. If you're having trouble finding a good "why" or "how" link, go back to your notes. If you still can't find one, check in with your study group.

There's one other technique that has proven effective in getting memories to stick: drawing a picture. It's not clear why it works, but it's probably due to the extra mental processing that is required to draw something. If I simply say to you, "Try to remember the word *potpourri*," there are few things you could think about to help you remember it. You might think of places you've seen potpourri—maybe in a display in a boutique around Christmastime—or you might think of the fact that potpourri is an unusually spelled word with that silent *t*. But to draw it, you must think of more. You must decide what ingredients will go into your potpourri and whether it's in a bowl or a basket, and those choices will probably make you think about what sort of room it's in. All of those details will help you remember the word later.

I don't recommend that you do this for everything you're supposed to remember, because it's too time-consuming. But **for material that just won't stick, try drawing a picture**.

> *In a sentence:* Meaningless material is hard to remember, so taking a little extra time to make it meaningful will probably be worth it.

TIP 35

Use Mnemonics for Meaningless Content Only

Occasionally you must memorize something that really is meaningless or close to it: the names of the twelve cranial nerves, for example, or rivers in Asia. In eleventh grade I was asked to learn the names of the US presidents in the order in which they had served. (I continue to be surprised by how often that information comes in handy.)

Mnemonics are memory tricks that help you learn something meaningless. One mnemonic technique requires that you memorize something simple, where the simple thing provides cues to the to-be-remembered content. For example, to remember the five Great Lakes, you memorize the cue "HOMES," which gives you the first letter of each: Huron, Ontario, Michigan, Erie, Superior. In other cases, instead of a word, you memorize a sentence, and the first letter of each word is a cue. Many medical students use the sentence "On Old Olympus's Towering Top, a Finn and German Viewed Some Hops" to remember the cranial nerves: olfactory, optic, oculomotor, trochlear, trigeminal, abducens, facial, auditory, glossopharyngeal, vagus, sensory, and hypoglossal.

Another mnemonic technique has you find ways to associate the facts you need to remember, one with another, often with visual images. For example, if you're trying to remember that the Spanish word for ribbon is *cinta*, since *cinta* sounds a little like "Santa," you might visualize a Santa with a bag containing ribbons instead of toys. Another technique relying on imagery is a mental walk. First, you must think of a mental walk or drive you might take—for example, from your home to a friend's house—and identify and memorize notable spots along the walk. The first notable spot on my walk might be my front porch, which is composed of an aggregate concrete I dislike. My second spot would be the stone wall halfway down my driveway that visitors keep hitting with their cars. Once you've got your mental walk memorized, you can learn a new, arbitrary list of objects by associating list items with the notable spots on your walk. For example, if you ask me to pick up bread, peanut butter, flour, and vitamins at the store, I could memorize the list using my walk. I would associate bread with the first notable spot on my walk (perhaps by mentally placing slices of bread to cover the unsightly aggregate on my porch), then I'd associate peanut butter with the second notable spot (perhaps imagining peanut butter instead of mortar used to repair my smashed rock wall), and so on. Later, when I need to recall the list, I go on my mental walk: I see my front porch, and I remember, "Right, I covered the porch with slices of bread. Bread was the first item on the list."

Mnemonic methods are often used by competitors in memory contests because they are asked to memorize things that have no intrinsic meaning, such as names to go with photographs of faces or the order of a deck of freshly shuffled cards. Memory contests use such materials exactly because they are lacking in meaning to all contestants, which makes them equally challenging for all. Meaning is helpful to memory, and what something means to you depends on what you already know about the subject. For example, it would be unfair to hold a memory contest in which competitors were asked to memorize a passage from F. Scott

Fitzgerald's novel *Tender Is the Night*, because some contestants might have read the book before.

There are many books on learning written by memory champions, and most of them emphasize the use of mnemonics, but mnemonics should really be your last resort. It's **a technique to be used only when you cannot make information meaningful**. That should happen rarely.

> *In a sentence:* Mnemonics help you memorize meaningless material, but they should be a last resort because it's better to make content meaningful.

How to Use Your Study Guide

All right, you've written your study guide, a comprehensive list of questions and answers. Now what?

You need to commit the answers to memory. That's fairly straightforward: ask yourself a question, and see if you can provide the answer without peeking. But there are a few tweaks you can add to this simple method to make it more effective.

First, **cover the answers from the start**. In other words, don't start by just reading the questions and answers, start by trying to answer the questions. Research shows that trying to answer questions even before you can know the answers adds a little boost to learning.

Second, it's a good idea to **speak aloud when you answer**. Again, there's research evidence that doing so improves learning. If you're in a place where speaking aloud would be awkward, whisper or speak subvocally. The reason this helps is not fully understood—it's not just the "out loud" part, because people remember lists of words better if they say them themselves compared to hearing someone else saying them. The source of the benefit might be that speaking out loud forces you to make your thoughts more complete.

Third, if the question has a longish answer (that is, it's one you've written to prepare for essays), you might **imagine that you're teaching someone else**. It's common knowledge that teaching others is a wonderful way to learn something, and this is an instance when research absolutely agrees with common knowledge. Remember, when you're quizzing yourself in this way, you may not be able to compose fully formed answers aloud. Rather, you'll be thinking more in outline form: "First I should talk about this, which raises *this* question, so then I'll talk about *that*."

Fourth, even if you're pretty sure you've answered correctly, **look at the answer you wrote for your study guide**. Immediate corrective feedback helps build the right memory if, by chance, you have the answer wrong. If you keep giving the same wrong answer to a question, it may help to explore why you keep making that mistake. Think about why that answer seems right and then explain to yourself (aloud) why the right answer is better.

Finally, **pose questions to yourself in random order**. Your flash cards will be lumped together by topic, because you wrote all the questions for a lecture or assigned reading at the same time. But test questions will probably not be lumped together by topic, and it's better to study them in the same way you will be tested. Also, if you test yourself with the same order of questions each time, there's some danger that your memory for answers will become tied to the question order; in other words, the answer to question 34 jogs your memory for the answer to question 35, but if someone asks number 35 after number 16, you won't know the answer.

Randomizing question order is easy on a digital platform or if you've used index cards that can be shuffled. If you wrote your study guide on a notepad, you can still bounce around the order of questions, but it's not ideal because it's hard to keep track of which questions you've asked. To me, this consideration is not important enough to dictate that you *must* compose your study guide digitally or on flash cards.

> *In a sentence:* Quizzing yourself with your study guide is straightforward, but your time can be made a little more effective with some techniques that ensure you don't breeze past the material but instead really think about it.

TIP 37

Don't Worry About Your Style

You may have wondered why I haven't said how studying should vary according to your learning style. After all, if everyone learns differently, how can I recommend the same strategies for everyone?

Scientists have conducted lots of experiments on this subject, and **the evidence shows no support for learning styles theories**.

Testing one of these theories is straightforward. Let's consider the most common learning styles theory, which says that people learn best either visually, auditorily, or kinesthetically (that is, via movement). An experiment would have three phases:

- **Phase 1:** Classify people as visual, auditory, or kinesthetic learners.
- **Phase 2:** Give people an experience according to one of the three styles. For example, some people see a series of drawings

that tell a story, some listen to a version of the story, and some (given some minimal instructions) act out the story. The crucial part is that for some of the people, their experience of the story matches their style, whereas for others, the experience does not.

- **Phase 3:** Test people's comprehension of the story, or perhaps wait awhile and test their memory of the story. We predict that when the story matches the person's style, they will learn better.

That's what we'd predict, but the data don't come out that way. People's supposed learning style doesn't affect their learning. There are at least fifty different learning styles theories, not just visual versus auditory versus kinesthetic but also linear versus holistic, visual versus verbal, and many others. There's no evidence that honoring people's learning styles helps them learn.

Despite the lack of evidence, the learning styles myth is resilient, and about 90 percent of the American public thinks it's backed by scientific evidence. I've written about it in several venues, so you can google my name and "learning styles" if you want to learn more.

> *In a sentence:* There's no scientific evidence for any learning styles theory, so don't worry about customizing your learning to your "style."

TIP 38

After You've Prepped on Your Own, Meet with Your Study Group

Although students are often encouraged to study together, research indicates that committing things to memory goes no better in a group. I think it's easiest on group members if you **meet to discuss what's likely to be on the test, then create and memorize study guides on your own, then meet again before the test**, maybe forty-eight hours before.

Why meet if you've already written your study guide and memorized it? This is where the differing perspectives of group members can prove helpful. For this session, I would recommend that the group **split into pairs and each member of a pair try to answer questions from the other person's study guide**. One benefit is that even though you've tried to make your study guide comprehensive, your group mates will have caught things you missed. A second benefit is that your group mates will phrase questions a little bit differently than you did.

One frustrating thing about memory is that it tends to be narrow when you learn something new. You learn a new concept phrased in a particular way. Even though there's nothing important about that phrasing, that's what sticks with you. So if everybody in your group uses their own language to describe the concepts you're supposed to understand, it will give you a broader perspective on what those concepts mean.

In a sentence: Meet with your study group after you've memorized your study guide to quiz one another; you'll each have slightly different perspectives, which will further aid your memory.

TIP 39

Remember That Cramming Usually Doesn't Pay

You've probably been told "Don't cram" since you first started taking tests. "Cramming" means budgeting most or all of your study time to fall very close to the exam. In other words, if you plan to spend five hours studying for an exam scheduled for Friday morning, cramming means studying five hours on Thursday night. An alternative would be studying one hour on each of the five days before the test—same total amount of study time, just distributed differently.

The change in timing has been studied by memory researchers for decades, and it makes a big difference to how well you remember content later. Here's a recent example that I especially like because it closely mimics the experiences of college students.

The researchers used subjects who were enrolled in an introductory psychology class. They picked sixty-four key concepts, then randomly selected thirty-two of those concepts for extra practice. They created a flash card slide deck for the items, and students worked with the deck until they got all the items correct. They had to do that three times, distributed across several weeks.

Then the experimenters analyzed the students' performance on the final examination, separating items that tested the practiced concepts from the other thirty-two concepts that didn't get extra practice but that students studied on their own. (Remember, this was a real course students were taking for a grade.)

When the researchers asked students how they had studied, they mostly said that they had studied the night before the exam—they had crammed. And the cramming kind of worked. Students didn't do terribly on the final exam for the "crammed" items—72 percent correct, compared to 84 percent correct for the items they had practiced with the flash card deck during the semester. So cramming doesn't lead to *great* performance, but it's not terrible.

What the researchers really wanted to know was whether students *continued* to show good memory for the information after the final exam. So they paid some students to return either three days or twenty-four days after the final examination to take another exam. The extra exam posed different questions that tested the same concepts.

After three days the students got only 27 percent of the "crammed" items correct. But for questions probing the practiced material, they got 80 percent correct. Even more amazing, the students who returned three weeks after that got 64 percent of the practiced content correct. **Cramming "works" so long as you don't care if you forget the information right after the exam. Distributed studying protects against this rapid forgetting.**

What should this result mean for you? There are circumstances in which I could understand why you might cram. Maybe you're taking a course just for fun and you don't care if you remember any of the content later—it's your lowest priority. That I understand. But think of how much extra work cramming creates if you later need to know the content of the course—for example, you take a more advanced course on the topic. If you plan to take Biology 102 and you cram for the final of Biology 101, you're just creating more work for yourself down the road.

One other thing you should know that's not obvious from the experiment I've described: **cramming *feels* as though it works well.** Imagine this: You and I are both trying to learn the sixty-four concepts from the introductory psychology course. I study the list for ten minutes on each of five nights. Each of those nights, when I start studying again, I've forgotten some of the content in the previous twenty-four hours. It's frustrating; it feels as though my studying is not going very well. But relearning is a great way to make memory durable.

Now suppose that, unlike me, you study fifty minutes on the last night. By the end of that fifty minutes, you feel great, as though you know the content cold. And in fact, the moment after we both finish studying that last night, you might remember more than I do. But two days later, you will have forgotten most of it and I will not.

The natural question is "If I don't put all of my studying into the night before the exam, how exactly am I supposed to distribute it?" People have tried to figure out the exact, maximally efficient practice schedule, and in fact there are apps that schedule practice for you at what is claimed to be the exact right time, depending on how long you hope to remember the content and how well you've done on memory tests so far.

I don't think it's worth worrying about what the best distribution of practice is. The main thing is to do some distribution of memorization. If shooting for the "perfect" distribution of practice means you're supposed to wake up at 5:57 a.m. on Sunday morning to quiz yourself on French verbs, you're just going to drop the whole thing. **Just do some distribution of practice and, if at all possible, have an overnight sleep between sessions.** In other words, it's better to study Tuesday evening and then Wednesday morning, rather than Wednesday morning and then Wednesday evening. Sleep is good for learning, as we'll see in chapter 11.

In a sentence: Cram only if you sincerely don't care about learning for the long term; otherwise, distribute your studying into multiple sessions across days.

TIP 40

To Prepare for Application Problems, Compare Examples

Instructors often like to include "application" problems on exams—problems that require you to use what you've learned, not just pull information from your memory. Because of the way human memory operates, these problems pose a special challenge.

You probably remember experiencing this challenge in school, especially in math class. For example, you learn about congruent shapes, and it all seems pretty straightforward, but then the exam has a word problem about diagonally cut sandwiches and napkins, and it doesn't occur to you that your knowledge of congruence applies. You learned about congruence with problems that used the word *congruence* and described simple geometric shapes. You'd get that sort of problem right, but when you read the test problem, your mind immediately goes to your knowledge about sandwiches and napkins, and that doesn't help you solve the problem. Later, you can't believe you failed to see what the problem was about.

We talked about memory cues in chapter 3. The specific aspects of the situation (here, sandwiches and napkins) are the memory cues, because the general principle that might apply (congruence) is hidden. It's not obvious whether the underlying principle of congruence applies to this problem, or calculation of area, or deductive logic, or what. But sandwiches and napkins are explicitly *in* the problem, so your mind treats *sandwich* and *napkin* as cues to memory and hunts for information connected to those.

The problem is not limited to math. For example, my students learn about Ivan Pavlov's famous experiment: the experimenter rang a bell, then fed a dog. With repetition, the dog came to salivate when it heard the bell, before it was fed. I expect students to recognize whether or not a very different situation—say, feeling anxious when they approach the classroom in which they failed a math test—is an example of the same type of learning.

How can you prepare for test questions that require you to apply what you've learned to new contexts?

One strategy is to **compare different examples of the principle you're studying**. In each of the examples above, something happens that leads to an automatic response that the learner can't help: the dog salivates when it's fed, and the student feels anxious when he's failing a test. Then something that doesn't prompt a response becomes associated with the thing that does prompt a response: the bell becomes associated with food, and the classroom becomes associated with struggling with math. With repetition, the once neutral thing (bell or classroom) begins to bring about the response (salivating or anxiety).

Comparing examples helps because it uses the memory-is-the-residue-of-thought principle. Comparing the problems prompts you to think about what they have in common, namely the shared general principle. Understanding the general principle is hard when it's stated in the abstract, but when it's described in the context of a concrete situation, it's easier.

> *In a sentence:* The best way to improve your ability to see the general principle in a problem is to find several examples of the principle and compare them.

TIP 41

To Prepare for Problem Variations, Label the Subgoals

We've looked at one variety of challenge in applying classroom knowledge to the real world: you look at a complex situation and fail to recognize "Oh, this is *that* sort of problem."

Other times, recognizing the problem is relatively easy, but the solution has a few possible variants, and what you learned was a set of steps specific to the example that the instructor provided. Consider this example, offered by the psychologist Richard Catrambone, of what is usually called a "work problem":

Tom can clean his garage in 2.5 hours. How long will it take him to finish cleaning it if his daughter already cleaned ⅓ of it?

Solution:

$$\left(\tfrac{1}{2.5} * h\right) + 0.33 = 1$$
$$(0.4 * h) + 0.33 = 1$$
$$0.4h = 0.67$$
$$h = 1.68 \text{ hrs,}$$

where h is the number of hours worked.

Based on this example, you might conclude, "The way to solve work problems is to divide 1 by one person's time, multiply it by the unknown, add what was already done, and set all of that equal to 1." That description fits the example. So far, so good.

But that set of steps doesn't apply for another work problem, even though it's similar:

Bill can paint a room in 3 hours, and Fred can paint it in 5 hours. How long will it take them if they work together?

Things went wrong because we described the first problem in very particular terms that applied only to that problem. We needed a set of steps that was just a bit more abstract: think about the amount of work done by each worker, then set it equal to the total amount of work to do. That is a conceptual description of subgoals:

First, I should represent the amount of work each worker does.
Second, I should set it equal to the total amount of work to do.

Labeling subgoals is a good way to ensure that you think about the general principle of multistep solutions to problems.

Here's another example. Suppose you're trying to learn how to use Gmail, and you look up the procedure to create a calendar event from an email. Tutorials commonly provide a series of steps that produce the desired outcome, such as this:

1. On your computer, go to Gmail.
2. Open the message from which you want to create an event.
3. On the ribbon of icons at the top, click the one with three dots.
4. Select "Create event."
5. Google Calendar will open, creating an event titled with the email subject line and inviting anyone else who received the email.
6. Set the date and time of the event.
7. Click "Save" at the top right of the screen.

This example could have the same drawback we saw with the work problems. Your understanding of the solution includes steps that apply only to this example, as it's specific to your computer and the procedure is slightly different on your phone. It's better to be mindful of the abstract principles that underlie the specific actions you're taking. For example. you could set them up like this:

Navigate to Message

1. On your phone, open Gmail.
2. Open the message from which you want to create an event.

Create Event

3. Tap the underlined date or time in the message.
4. Choose "Create event" from the menu that appears.

Complete Missing Properties

5. A window opens, creating an event titled with the email subject line and with the date and time from the email.
6. Change the duration if needed. (Default duration is one hour.)
7. Invite others to the event if needed.

Save Event

8. Click "Add" at the top right of the screen.

Labeling the subgoals may seem like a trivial change. But the labels facilitate two mental processes that we've already seen can aid learning: first, **they make explicit the organization of the steps,** and second, **they emphasize meaning**—they make clear *why* you're carrying out each step.

To use this strategy, start with an example provided in your textbook that's meant to illustrate a particular type of problem solution. Scan the accompanying text that describes the procedure in more abstract terms.

Then match the two by creating labels for the subgoals that the concrete steps are meant to address. If possible, find someone who understands the procedure well to give you feedback on whether you've gotten it right.

In a sentence: When you learn a multistep solution to a problem, part of the solution may be specific to that one problem; to help you apply your knowledge more broadly, try labeling the subparts of the solution.

For Instructors

Naturally, it's up to learners to implement the study methods described in this chapter, but there are a few things instructors can do to smooth the way.

One is to be forthcoming about what will be tested and how. Because they are new to the field, students are not good judges of what is really central information and what's a fun fact you just threw in. Tell them.

You can also provide guidance about what I've called "found materials." Mimic the search a student would do when looking for study materials for your class. Then tell your class about the quality of what you found and remind students why creating their own materials leads to greater success in the course.

When it comes to getting learners to use the learning tips described here, certainly you can simply tell them, "These strategies are good, and these others that you might be using are less effective," but to avoid overwhelming them, you might start with the three learning strategies you think have the greatest chance of being used.

It might be even more effective to demonstrate the techniques to students rather than telling them. You can:

- Use retrieval practice by giving low- or no-stakes quizzes in class.
- Use distributed practice (see tip 39) by revisiting content at targeted times of the marking period.
- Use the power of meaning (see tip 34) by emphasizing to students the links among what seem like disconnected facts.
- Describe to students what their study guide should look like (see tip 32). Devote a little time at the end of a few classes to letting them practice writing study guide questions for the material covered that day. Provide models and have students share the questions and answers they compose.

Just deploying these learning tips in class will result in better learning, but you can also go all out by repeating the experiment I described as part of tip 39. Employ a learning tip in class for some of the content but not the rest. Then, on a unit test, separate students' performance on the content you targeted. Show students how much better they did on that content, explain why, and then emphasize that they can do this kind of work as part of their own studying.

Summary for Instructors

- Tell students what information they are and are not expected to memorize for exams.
- Talk to students about the value and reliability of "found materials."
- Advise students how to study.
- Put principles such as distributed practice and retrieval practice to work during class time.

CHAPTER 7

How to Judge Whether You're Ready for an Exam

A lawyer preparing for the bar exam doesn't set a time—say, one hundred hours—that she'll study and then stick to that. She evaluates her learning as she goes and stops when she thinks she's mastered the material. Thus, learners must be confident that their judgments about what they know are accurate.

You've surely had the experience of thinking you were ready for a test and then somehow doing poorly anyway. People in this situation often blame the test. They figure, "I *know* I knew the content. Therefore, there must be something wrong with the test because it did not show that I knew it." But your judgment "I knew the content" is the result of a mental assessment. Maybe *that* was the lousy test, not the exam the instructor administered. It may come as a surprise, but people can be mistaken about what they know.

Many Factors Contribute to Judgments of Learning

Suppose you are taking a conservation biology course and you want to cement this fact in your memory: the red-handed howler monkey is native to Brazil. How would you know whether you've learned it? Easy: ask yourself, "The red-handed howler monkey is native to which country?" and see what pops out of your memory. Certainly, that's one way to judge whether you know something, and it's a good one.

But **people confuse performance and learning**. Here's the difference. Suppose I see you right after a workout, and you tell me you've been practicing push-ups and can do twenty. I say, "Cool, show me!" You might say, "I can't *now*; I'm tired from my workout." You've learned to do twenty push-ups, but your performance wouldn't show that learning in the current circumstances.

When it comes to learning, "performance" means saying "Brazil" in response to the question "The red-handed howler is native to which country?" You can see why you would think, "I answered the question, so I definitely know that one." But the fact that you can answer it now (under one set of conditions) doesn't mean that you will be able to access that memory reliably under all circumstances.

For example, you might have learned to speak conversational Japanese pretty well, but your performance with a Japanese border agent doesn't show your learning because you are tired from your flight and a little nervous. (Or maybe it's just me who gets nervous for no reason when talking to border agents.)

Typically people overestimate what they know because they test their knowledge in ways that, without their realizing it, support their performance. Thus, they judge that they have learned something because their performance is good when they quiz themselves, but in fact their memory is shaky.

WHEN JUDGING WHETHER YOU'VE LEARNED SOMETHING
What your brain will do: It will confuse performance and learning. If you recite something from memory—even though you aren't really drawing on your memory—your brain will conclude that you've studied enough.
How to outsmart your brain: Test your knowledge without any other support to your performance. The easiest way to do that is to mimic the conditions of an exam.

In this chapter we'll look at three ways that people can be deceived about their learning when testing themselves, and I'll describe self-tests you can do to get better information about what you really know.

TIP 42

Be Clear About What It Means to "Know" Something

In his *Confessions*, written in about the year 400, Saint Augustine noted, "If no one questions me, I know; if I want to explain it to those who question me, I don't know."

This distinction is timeless. Every instructor has had a conversation with a student that went something like this:

Student: I can't understand how I could have gotten such a low grade. I studied so hard, and I know that I knew everything! Some of the questions seemed really ambiguous to me.

Instructor: But you knew everything. . . .

Student: Yes!

Instructor: So, for example, you'd be comfortable explaining the different mechanisms of forgetting.

Student: For sure.

Instructor: Okay, so why don't you describe the main theories of forgetting we talked about.

Student: Okay. There's a stimulus and a response. And if the stimulus is not connected to the response . . . wait . . . no . . . yes, that's right, if the stimulus gets cut off from the response, or wait, not cut off . . . um . . . well . . . I know it, I just can't explain it.

This student is using the word *know* differently from the way instructors do. The student is thinking, "When we first started studying how forgetting works, it made no sense to me. I didn't understand the textbook chapter, and I didn't understand the lecture. But I went over the reading really carefully, and a friend from class explained some of the concepts in a different way, and now when I hear the theories of forgetting, it all makes perfect sense."

You can see why the student feels he understands; he *is* much farther along than he was. Following along when someone else discusses an idea is partway to the understanding that instructors expect. But it's not enough. **Being ready for a test means being able to explain content yourself, not just understanding it when someone else explains it.**

This situation is a good example of the difference between performance and learning. My student is noting his performance: "I'm following this discussion really well, and a few days ago it would have been really confusing!" He's not considering that this performance does not necessarily signify complete learning.

Unfortunately, the way that many people study leads them to exactly this mistaken perception of what they know. Let's see how that happens.

In a sentence: "Knowing" doesn't mean being able to understand an explanation; it means being able to explain to others.

TIP 43

Rereading Leads to Overconfidence in Your Knowledge

Imagine you're taking a business school course called Innovation. You attend a lecture on wearable technology: clothing and jewelry that collect and store physical information such as heart rate and body temperature. It's pretty interesting, and you have little trouble following it. The next time the class meets, the professor starts to deliver exactly the same lecture. A nervous titter runs through the room, which the professor ignores. Soon, someone raises their hand and points out that he already gave this lecture. The professor says, "Yes, but it's important material, so it's worth repeating." He proceeds to give a presentation identical to the previous one: same slide deck, same anecdotes, same "spontaneous" jokes.

What would you think?

If you're like me, you would think it was a big waste of your time. I'd be thinking, "Yes, yes, you said that last time. I know all this, I'm not learning anything."

Now, do I *know* the content the speaker is reviewing? Yes and no.

On the one hand, I know I have heard it before, and that judgment is based on my memory of the previous lecture. In that sense I "know" it. But if I tried to provide a summary of what he said, it wouldn't be very good.

Many memory researchers distinguish between two ways that you can pull information out of memory. One method is rapid and requires very little attention, but it can provide only limited information; it identifies whether or not something is *familiar*. It tells you whether you've encountered something before but not anything associated with it, nor where or when you encountered it. Another memory process can provide information associated with something, but this process requires attention, and it occurs more slowly.

These two types of memory probably ring a bell. Sometimes you'll see someone on the street and the familiarity process tells you, "You know this person!" So you call on the other process for more information: What is this person's name, and how do I know them? That second process may deliver nothing—you have no information about their name, how you know them, or anything else. That doesn't make you feel less certain that you've seen them before.

In tip 30 I mentioned that rereading is one of the most common study techniques and pointed out that it's not an effective way to commit something to memory; you should think about meaning, and rereading doesn't guarantee that.

Here we consider another reason why rereading is a bad idea: **rereading misleads you into thinking, "I know this."** Rereading is like going to the lecture on wearable tech for the second time. When you're rereading, you're thinking, "Yes, yes, I've seen all this before. This is totally familiar." But that's just it—the sense of "knowing" you're getting is from the memory process that assesses whether or not you've seen something before. You're right, you have seen it before, but knowing you've seen it before is not the same as being able to talk about or analyze the content. And the more you reread, the more the process that assesses familiarity tells you, "You've seen this before!"

To be clear, **rereading is desirable for the purpose of comprehension.** If you read something and didn't understand it, give it another try. But rereading is a bad way to commit something to memory. It's bad enough that it doesn't help memory much, but in addition it makes you believe that your knowledge of the content is improving.

So what can you do to get a more accurate assessment of how your studying is going?

In a sentence: Rereading boosts familiarity, giving you a false sense that you have mastered content, but being familiar with something doesn't mean you can recall it from memory and provide other, related information, which is exactly what you need to do for an examination.

TIP 44

Evaluate Your Preparation with the Right Type of Self-Testing

At the start of this chapter I noted that the question "How do you know whether you know?" sounds stupid because it seems easy to evaluate: you look into your memory and see if the information is in there. The problem, as we've discussed, is that you might rely on *other* information (usually, a feeling of familiarity) to make that judgment. Your first instinct—to look into your memory—was good, but **you need to ensure that you actually test your memory for the content.**

People understand that part of exam preparation is self-testing. The mistake they make is not challenging their memory the way it will be challenged on the test. They read over their textbook and then look up and try to summarize the section they just read. If they can provide a good summary, they figure that's evidence that they've mastered the content. This is a decent way to test comprehension—if you can paraphrase what you just read, you understand it. (It's not foolproof, however, because you have no way of knowing whether your paraphrase is accurate.) But if you aim to commit something to memory, this self-testing is faulty in three ways.

First, you can't self-test for material you just finished reading. You're not *really* testing your memory, because the content is still rattling around in your short-term memory—you just read it! There's no hard-and-fast rule here, but I'd say **at least thirty minutes should elapse between when you read content and when you self-test on that content**.

Second, summarizing is fine as one type of self-test, but you really want to quiz yourself on other content: knowledge of specific details, inferences you can draw, comparisons among ideas, and so on.

Third, when you self-test, you need to articulate your answers aloud. When you answer in your head, it's too easy to be satisfied with a vague or incomplete thought. Answering aloud and articulating full thoughts makes it plain to you whether you've really got it.

You probably noticed that the conditions I set for self-testing—test when you haven't seen the answer recently, use varied questions, say the answer aloud, get feedback—is pretty much baked into the procedure I encouraged you to use for studying in chapter 6. **If you write a comprehensive question-and-answer study guide and study by testing yourself, you will get good information along the way regarding how much you know.**

In a sentence: To assess whether you really know something, you should test yourself when you haven't seen the content recently and say the answers aloud—a practice that pairs easily with the method I suggested you use to memorize your study guide.

TIP 45

Don't Use Practice Tests to Judge Your Readiness for an Exam

I've mentioned that my students are eager to get their hands on exams from prior years. But I actually think old exams do more harm than good, at least the way my students use them.

Even if they know that the questions will be different this year, students use old exams to judge whether or not they are prepared. The logic seems to make sense: if I take last year's exam and score 90 percent, it would seem that I'm ready for this year's exam.

There are a couple of reasons **you shouldn't use previous tests to evaluate your readiness**. First, last year's test probably doesn't exactly reflect this year's content. The readings may have changed, the pace of the course might be faster or slower, some topics have received greater or lesser emphasis, and some content may have been updated. These changes accumulate as time passes, so if a student collects exams from the last several years (probably gloating over her prize), the older ones will reflect this year's course all the more poorly.

Naturally, students have no way of knowing how the course has changed. So if the term *blocking* appears on a test from two years ago and I didn't discuss blocking this year, students with that old exam will panic.

"Why can't I find *blocking* in the book or my notes?" That problem is unfortunate because of the panic, but at least it's easily fixed: the students ask me about blocking, and I tell them we're not covering it this year. (Actually, I usually explain blocking first and *then* tell them that it's not part of the class and so won't be tested. This practice does not make me more popular.)

There's a more serious drawback to using old tests to evaluate your preparedness. Suppose there were, say, a thousand concepts covered in a course. Naturally, a test won't include questions on every concept, so you might say to yourself, "I think I'll learn nine hundred of them, and maybe I'll be lucky and the instructor won't ask anything about the hundred I've ignored." That's a bad strategy, because there's no reason you should expect to be lucky. Ideally you want to learn everything. Even with that goal, there will be some things you know better than others and therefore some element of luck in your exam score, depending on what happens to appear on the exam.

It's obvious, then, that if you want to minimize chance, **you should judge whether you're ready for the test based on how well you know** *everything*. If you take an old test to decide if you're ready, you are judging your readiness on just a fraction of the content to be learned. You're throwing an element of chance into your preparation when you don't have to.

It's smart to **look at old tests to get a sense of the *types* of questions that an instructor tends to ask**. But don't use old tests to measure whether you've done enough studying. Judge whether you're ready for a test by how well you know the content of the study guide you've written.

In a sentence: Use old exams to get a sense of the types of questions that might be posed, not to judge whether or not you have done enough studying.

TIP 46

Study Until You Know It; Then Keep Studying.

Let's try another thought experiment. Suppose you're taking a world history class and the instructor announces that there's a quiz on Monday. You're to know the names and dates of the ancient and imperial Chinese dynasties, but there are only sixteen, so it doesn't seem too challenging. You quiz yourself Sunday night until you are able to recite the list in order perfectly.

Will you remember the names and dates for the quiz the next day?

You might think, "I tested myself and knew them. So obviously I know them. What's the question?"

You might remember the dynasties perfectly the next day, but you probably won't. Eighteen hours or so will elapse, so you'll forget some of what you studied. Remember that I drew a distinction between *learning* and *performance*? People tend to think that their performance at any given moment reflects their long-term stable learning—if my performance is 100 percent today, it will be 100 percent tomorrow, because 100 percent reveals the state of my knowledge.

The only way to address the problem is to anticipate forgetting. You need to **study until you know it, and then keep studying**. This practice, called *overlearning*, has been examined extensively in laboratory experiments, and there are two things you should know about the research. First, overlearning works, just as you would expect it to. It protects against forgetting. Second, while you're doing it, **it feels as though it's not working**. It feels pointless, even foolish, to keep studying after you know something. You're going through your flash card deck and getting every answer right, so you can't help but wonder, "What good is this doing?" What it's doing is strengthening the memories to shield them from forgetting.

How much overlearning should you do? It depends on how long you hope to remember the information, the nature of the content, what else you know about the topic, and other factors. When I was in college, I remember talking with a friend during finals week about her preparation for her organic chemistry final. She told me, "When leaves blowing around on the quad look like organic compounds to me, I know I'm ready."

That seemed a little intense to me at the time, and it still does. As a rule of thumb, I'd advise studying until you know the content, then add about another 15 percent of study time. There's nothing especially research based about that number; the important thing is that you do *some* overlearning, whatever the exact amount ends up being.

In a sentence: Don't study until you know the content and then stop; keep studying a little longer to protect against the forgetting that will happen during the time between when you stop studying and when you take the test.

For Instructors

I've reviewed two key ideas in this chapter: (1) there are different ways one can "know" something, and (2) our judgment of whether or not we know something can be faulty. Both ideas are useful for students to understand, and both can be brought home to them via in-class demonstrations.

Here's one to help students understand that there can be different ways to know things. First, ask students to list, on a blank sheet of paper, as many US states as they can in, say, three minutes. Depending on their age, students might list twenty or thirty. Next, give each student a map of the United States showing nothing but outlines of the states. (For students outside the United States, use local geography, as appropriate.)

Everyone will be able to name many more states when looking at the map. Rereading your textbook is like looking over the map and thinking, "I can name all the states." The test, however, is akin to naming the states in the absence of the map.

What about helping students understand that their sense of "knowing" can be faulty? I find in-class opportunities to highlight this difference almost any time I explain something complicated. If I then call for questions (with a good long wait time) and get none, I'll say, "Okay, turn to your neighbor and take turns explaining to each other those two theories I just went over." Invariably, many students quickly realize that they can't do it. I then explain what just happened: they evaluated their knowledge not by their ability to explain but by their level of understanding when someone else explains.

I find this method more effective than pop quizzes, by the way. When students can't explain something on a quiz, they don't conclude "I guess I didn't know it after all." They conclude the quiz was not fair.

Once your students understand that there are different ways to know things and that their feeling of knowing can mislead them, be sure to connect the dots for them and tell them exactly what it means for their own learning, namely that they must keep in mind how they will be asked to demonstrate their knowledge on exams and they must use reliable methods to assess their learning when preparing. Then tell them how to do that.

Knowing that there are different types of memory should also make you more likely to tell your students what to expect on a test. As an instructor, I'm tempted to tell students nothing about upcoming tests: I want to say, "Just know the content and you'll do fine." But as a memory researcher, I know that is a bit idealistic.

Because recognizing is much easier than recalling, multiple-choice tests, which show the answer, usually demand quite fine-grained knowledge. Short-answer tests seldom require such detail, but students don't have the benefit of seeing the choices laid out for them. And of course for an essay test they need knowledge of broad themes.

It's asking a lot of students to have the depth of knowledge needed to answer questions in any format on any of the class topics. I think it's wiser to set realistic expectations for what they should know and tell them what your expectations are.

The students who struggle the most on exams seem to have the most trouble evaluating whether or not they are ready. Helping students understand the difficulty of assessing their knowledge may take some persistence. But it's worth your effort, because a more accurate sense of what they know will enable them to prepare more effectively.

Summary for Instructors

- Conduct an in-class demonstration so students understand the difference between understanding when someone else explains versus explaining themselves.
- Show students that their judgment of whether something is in their memory can be faulty.
- Tell them how to self-test.
- Let students know what type of knowledge you expect on tests.

CHAPTER 8

How to Take Tests

Exams require two things: that you recall information from memory and that you do something with that information, for example, solve a problem or write a persuasive essay.

I've noticed that students change their judgment about which of these requirements is more important, depending on where they are in the test-taking process. Before a test, they think that their success will be determined by how much stuff they have in memory: "If I study, I'll pass." After the test, they think the same way: they think their grade was mostly determined by how much they studied.

During the test, however, students think very little about the process of getting information out of memory and focus on how to most effectively use whatever memories they can access. They spend a lot of time and mental energy trying to interpret *what questions really mean* or to guess *what the instructor wants me to say.*

These are commonly called "test-taking strategies," but they often go wrong. They prompt students to interpret questions as having subtle meanings other than what's plainly asked. Or students try to eliminate answers to a multiple-choice question based on supposedly helpful tricks such as "Answers containing the word 'always' or 'never' are usually wrong."

Students use these dubious strategies during a test because they're trying to get maximum value from the first information they can pull out of memory. But they seldom try to get *more* out of memory during a test because they think there's no point. But they're wrong.

In chapter 3 we saw that a memory can be "in there" but difficult to retrieve because of the way that you probe your memory. (Seeing the grocery store was not a good cue for me to remember to buy coffee for my neighbor's gift basket.) With the right strategy, you might be able to coax a reluctant memory out of the vault.

WHEN TAKING A TEST

What your brain will do: It may believe that if you know something, each attempt to retrieve the memory will be successful. Actually, *working* at remembering something can pay off. But instead of trying to squeeze more out of memory, people apply ineffective strategies to the content that comes out of memory easily.

How to outsmart your brain: Don't give up on your memory if it doesn't provide the desired answer right away. Test-taking strategies should be your very last resort.

In this chapter we'll look at a few ways to create cues for memories that you can't recall. I'll also elaborate on why it's usually a bad idea to use test-taking strategies or tricks. But strategies meant to keep you organized and calm are useful, and we'll consider those first.

TIP 47

Prepare and Take Care

What's more frustrating than getting a test back and seeing an answer marked wrong for which you absolutely knew the content but did something dumb, such as circling *b* when you meant to circle *c*? To prevent this problem, you need to add a few simple routines to your test taking.

Routine 1: Spend the first thirty seconds or so reading the instructions, if any are provided. Usually they don't say anything important, but sometimes you learn that there will be a penalty for guessing, for example, or that you needn't show all your work, or that you needn't write in full sentences.

Routine 2: Spend the second thirty seconds skimming the test to get a sense of how much time you can spend on each question. Pay attention to the point values of different questions so you can allocate more time to questions that count for a lot. Do a quick calculation that tells you where you should be when one-third of the time has elapsed and when two-thirds of the time has elapsed. When making these calculations, don't forget to leave a little time at the end to check your work. Mark those spots on the exam.

Routine 3: Read. Each question. Carefully. My students sometimes get questions wrong because they read half of a question and assume they know what it's asking. Or they read the whole thing but somehow do not see the word *not*. When you think you have the answer to a question, take an extra few seconds to be sure the question asked what you thought it did.

Routine 4: In the last few minutes, check your work. Be sure you didn't unintentionally skip questions. Read over your essays for illegible or missing words, unfinished sentences, and the like. For multiple-choice questions, be sure you ticked the box that you intended to. For math or science tests, make sure you actually completed multistep problems. Circle your answer to each question so it's clear to the grader what you mean to be the solution. Be sure that units are specified. Label the axes of graphs.

If a student comes to see me, frustrated about a low exam score, there are almost always a few percentage points that end up being attributed to "stupid mistakes." Make these routines into habits, and you won't lose points for silly reasons.

> *In a sentence:* A small set of routines to help with planning and attention to your work will ensure that you don't lose points due to carelessness.

TIP 48

Learn to Cope with Ordinary Test Anxiety

Most everyone gets jittery during a test. Most people also get nervous when they speak in public or attend a social event at which they don't know anyone. The fact that it's typical doesn't mean you can't learn some ways to handle it. Nerves are distracting, and they will affect your exam performance. (If you have a lot of trouble controlling your anxiety or if you get anxious not just when taking exams but when preparing for them, see chapter 14.)

As you no doubt know, anxiety is self-perpetuating: anxiety during a test makes it hard to concentrate, which makes it hard to remember answers, which increases your anxiety. Thus, strategies to control test anxiety focus on breaking this cycle or never entering it in the first place. Here are some techniques you can try.

Try reducing your consumption of caffeinated drinks on exam day and see if that helps.

Some people are made anxious by the presence of other test takers—just seeing someone else nervously jiggling his foot or, worse, confidently racing through the exam sets some people's hearts racing. If you're one of those, **try isolating yourself**: sit alone or near the front of the room if you can. Or you can try wearing earplugs during the exam. It makes some people feel as if they are in a world of their own. (Check with the instructor to be sure this is okay.)

Some people calm their anxiety by venting about how anxious they are or how high the stakes are for this exam; it makes them feel better, but it can be terrible to listen to, so you might **avoid chatting with other test takers** right before the exam. If the venter is a friend and it would be awkward to shut down his conversation, tell him you want to do some last-minute cramming, and keep your nose in your notebook. (If actually reading your notes focuses your mind, great; if it just adds to your nerves, you don't actually have to read.)

Some people like to **meditate or pray** before an exam to calm their mind and feel centered, and it's a good strategy to try if you start to panic during an exam. If you do neither practice regularly, here's a simple three-step procedure that will serve: (1) Close your eyes. (2) Breathe in slowly to the count of seven and exhale to the count of seven. (3) Repeat two or three more times or until you feel your body begin to relax. If that doesn't do the trick and if bathroom breaks are allowed, taking a brief walk sometimes provides a mental reset.

Sometimes it's hard to relax because your thoughts are running away from you. Some **realistic self-talk** after this breathing exercise might help. If you're panicking because the test has a bunch of questions you didn't expect to see, remember that everyone else is in the same boat. If you're panicking because you didn't prepare, remember, this one test doesn't determine your future, much less *you*. If you fail, you can bounce back. Have confidence that you can make a plan for bouncing back later—perhaps think of someone who could help you formulate the plan. Promising yourself that you'll make a plan and assuring yourself that people will assist you might allow you to set your panic aside and try your best to do the task before you.

I've sometimes heard advice about dealing with anxiety by "visualizing yourself succeeding." I always found that a little hard to maintain. For example, I might be on a panel of speakers, and I'm saying very little, but every time I open my mouth, people seem to look at me as though I've said something silly. So now I'm just keeping quiet and not contributing anything. At this point I could try to visualize myself saying really smart things and people dropping their jaws in delighted wonder, but I simply wouldn't believe my visualization.

If visualizing success works for you, great, but if it doesn't, here's an alternative: **visualize someone supportive being with you**. I can't visualize success at this terrible panel, but I can imagine my wife by my side, and that helps in a few ways. It makes me see my performance through her eyes, and her perception is more realistic than mine. Okay, maybe I'm *not* at my best right now, but it's not the horror show I'm making it out to be, either. That said, I *should be* trying my hardest instead of dwelling on my performance and saying nothing; I owe that to the people who invited me. I can also hear in my head supportive things my wife would say about how it's going: "Some people did seem interested when you spoke, like that woman in the third row, she was nodding. And another thing, what's up with that other dude on the panel, the one at the end of the table? Why is he talking so much? You're definitely more interesting than he is."

The next time you get nervous during an exam, imagine someone supportive sitting there with you, the sort of person who always knows what to say to make you feel more confident.

We've discussed how to refrain from making careless errors and how to keep anxiety under control—in short, how to keep things from going wrong. What about methods for making your test performance better? Let's turn to some techniques that will enable you to induce reluctant memories to come out of the vault.

In a sentence: You can combat mild test anxiety by avoiding situations that make you anxious during the test and by using self-calming techniques when you feel stressed.

TIP 49

Imagine Yourself in the Place Where You Studied

If you're a student taking a written exam, you are likely in the same room where you learned some of the content—the classroom—but of course you probably didn't prepare for the exam in the classroom. And if you're taking a police officer entry exam, for example, or a teacher certification test, you're probably in a room you've never seen before. Is your memory worse when you try to recall something in a location different from the one where you studied?

Maybe a tiny bit.

Here's why. Sometimes it's important to know *where* and *when* something happened; that's usually called contextual information. For example, when I park my car at the grocery store, I want to remember where it is, but I don't want my car *permanently* associated with that particular spot. I want it associated with both the spot *and* a time: I parked in this spot today, not forever. That's why finding your car can be so hard. You're trying to find the memory time-stamped "today," but it's easily confused with the other memories of parking in the same lot (or similar lots) time-stamped with different days.

When you're studying, you don't want your memories to be tied to a context. You may have been in your kitchen when you learned that Wilhelm Wundt is considered to be the first modern experimental psychologist, but you don't want "kitchen" tangled up with that memory; he was still the first modern experimental psychologist when you're in your living room or the classroom.

Unfortunately, time or location can enter into a memory when you don't want it to. Researchers test this idea with a simple experiment. Subjects are told that their memory will be tested, and all they need to do is listen carefully to a list of words that will be read aloud. The experiment takes place in a tidy classroom with a big window, and the words are read by a young man who is dressed rather sloppily. Two days later people return and try to remember the words. For some people the test is administered by the same young man in the same room, but others take the test in a small, cluttered office, administered by a smartly dressed older woman. Those people don't do quite as well on the test.

Changing locations doesn't have a *huge* effect on memory. Memory wouldn't work very well if it did. Imagine going out for a lovely meal with your family to celebrate a birthday and being unable to remember it unless you went back to the restaurant. But of course you'd like to account for the effect it may have and avoid any detriment to your learning.

Learning is much more likely to be tied to a place and time if you think about the place or time as you learn, for example, if you use the cracks on your living room wall to help you visualize the major rivers of Western Europe. It's another instance of the principle that memory is the residue of thought. So as you study, you want to avoid any conscious tying of the content to the study environment.

Nevertheless, the environment can still creep into your memory even if you don't consciously include it in your studying. To use this possibility to your advantage, if you're in a test and are having trouble recalling something, **try visualizing the place where you studied it**. Imagine yourself in that location. If there were characteristic sounds or smells there, put those into your imagining as well. This visualization may help you recover the lost memory.

In a sentence: If you're having trouble recalling a fact you've studied, try visualizing the place where you studied it.

TIP 50

If You Can't Remember a Fact, Think About Themes

Let's try another little experiment: see how many animals you can name in sixty seconds. Really, give it a try. (If you feel as though you can't name any more after a shorter time, you can stop.)

How'd you do? Suppose I gave you a hint, such as "Animals you see on a farm." Can you name any new ones? How about "Animals you'd see in Australia"? Or "Circus animals." Or "Animals you see at a pet store."

Memories tend to be organized in themes, or clumps, and they can be retrieved that way. An experiment showing this principle was conducted by the psychologists James Pichert and Richard Anderson in 1977. They had subjects read a brief description of two boys, Mark and Pete, skipping school and hanging out at Pete's house. Everyone read the same passage, but some were told to read it as though they were a burglar and others as though they were a home buyer. Later, the subjects' memories of the passage were tested, and those who had read the story from the burglar's perspective remembered things such as Pete saying that the side door to the house was always left unlocked and that his father had a rare coin collection. Those taking the perspective of a home buyer remembered that the stone siding was new but the roof was leaky. This effect is pretty easy to understand; if you're told to think like a burglar, as you're reading you notice details that would be of interest to a burglar. Likewise, the home buyer's perspective prompts you to notice details relevant to that.

But here's where things get interesting: if the experimenters then asked people to switch perspectives—the burglars were asked to think like home buyers and vice versa—they remembered information that was relevant from the other perspective. When the burglars were asked to think like home buyers, they presumably thought, "Hmmm. What would a home buyer care about? A nice neighborhood? Whether the house is in good repair? Oh, right, the story said something about a leaky roof."

How can you use this principle when you're taking a test? I doubt that it will help you recall a forgotten nugget of information when it's very clear what's being asked, for example, "In what year was the Treaty of Versailles signed?" But thinking of broad themes could be useful for the integrative questions you often find on essay exams, for example, "What were the most important consequences of the Treaty of Versailles for France during the 1920s?" Broad themes can also help you think through questions asking you to apply what you know to a concrete situation, for example, "Outline an approach to developing a virtual reality environment to simulate a roller-coaster ride." In these types of questions, you're given limited clues about which part of the vast amount of information you've learned is relevant to the answer. You're forced to do an undirected search of memory, just as you did when I said, "See how many animals you can name in sixty seconds." You might find yourself thinking, "Did I learn anything about this?" Or perhaps you remember a few things that seem somewhat related but don't seem very promising as the start of an essay.

If you're having a hard time figuring out which part of what you've learned is relevant to a question, **list the topics you've learned on a piece of scrap paper** or the margin of the test. This will serve the same purpose as my hints: Australian animals, farm animals, and so on. For the question about the Treaty of Versailles, you might list "Financial impact of the war," "Territory gained because of the treaty," "Social reintegration of soldiers," and so on. Once you've listed as many topics as you can, go through them one by one to see if they jog any memories that help you interpret and answer the question.

Needless to say, this process may take a while. You should do it only when you've been through the rest of the test and have time to return to the problem that has stumped you.

In a sentence: Some test questions provide only very general cues to memory, and you may not consider one or more broad course topics that are relevant while you're formulating an answer; in that case, list the themes of the material you've covered to be sure you consider all the content that might be pertinent to a question.

TIP 51

Keep Trying

Here's another thought experiment for you: Suppose you agree to participate in a memory experiment: I show you forty-four line drawings of common objects (a fish, a flower, and so on), each one for five seconds. You see the whole set twice. Twenty-four hours later, you return and I give you a blank sheet of paper, asking you to list as many of the objects depicted as you can in five minutes. Then for three minutes I ask you to do an unrelated task (simple math problems) so that you don't think about the drawings. Then I give you another blank sheet of paper and again ask you to list as many of the objects as you can remember. Then you do another three minutes of math problems and then one more test of your memory of the line drawings you saw the day before.

How do you think you would do on the first, second, and third attempts? Better? Worse? The same?

In this experiment, people remembered an average of nineteen objects on their first attempt, a little over twenty on their second, and twenty-one on their third. This general result is typical: **people remember a little more each time they attempt to remember.**

This phenomenon has been observed in many experiments over several decades, but why it happens is not completely clear. At least part of the effect is due to retrieval practice: looking for something in memory makes it more memorable, even if you don't find it (see chapter 6).

You may also remember a little more because the same cue can work slightly differently at another time. Picture making the break shot in a game of pool; although the balls are always configured in the same triangle and they *look* as though they are positioned in just the same way each time, there are actually subtle differences, so that even if the cue ball hits them in just the same way, the outcome is different. Sending a question into memory such as "The Battle of Karbala was crucial in splitting which religion into two factions?" and seeing what comes out is a bit like hitting the cue ball and seeing where the other fifteen balls go. Even though your mind *seems* as though it's the same every time you pose the question, it may be a little different. And one time that difference means that the cue will produce the desired answer.

Now, there's no point in reading a question, failing to come up with the answer, and then immediately rereading it, because your memory really is the same as it was five seconds ago. But if you return to the question in five or ten minutes, your memory will be in a slightly different state because you will have been thinking about other questions. And so it may respond differently to the Battle of Karbala question.

When taking a test, for each question, **try to remember the answer for thirty seconds or so. If that doesn't do it, mark the question and come back to it in five or ten minutes.** Keep at it until you run out of time or finish the test.

That advice raises the question of whether it's smart to change an answer or whether you're better off sticking with the answer you first thought was right. Researchers have examined this question in many experiments going back to at least the 1960s, often using the same technique: they check exam papers for erasure marks and categorize each change as (1) a wrong answer to a right one, (2) a right answer to a wrong one, or (3) a wrong answer to another wrong answer. They consistently find that students mostly change wrong answers to right ones. Furthermore, when asked about the changes they made, students rarely say that they were due to realizing that they marked the wrong choice by accident. Instead, most changes are due to continued thought about the question; the test takers keep trying to remember, and it pays off with a new insight or inference.

For example, suppose you've answered "Buddhism" to the Battle of Karbala question, which you know was sort of a guess—"Buddhism" just came to mind when you saw "Karbala." But while answering another question you remember the instructor mentioning that many observant Shia Muslims wear black to mourn someone whose name you can't remember, but you're pretty sure that person was the martyr of the Battle of Karbala. So now you're 95 percent sure that the answer is "Islam," not "Buddhism." Clearly, you should change your answer.

But what if you're not that sure? This situation is especially common in multiple-choice exams, where questions are written to have more than one answer choice that looks right. If you are stuck between two answers that look equally good, should you go with your first instinct or your second? I have not seen an experiment that answers this finer-grained question, and to be honest, I'm not sure there is a general answer that's right for everyone. I think you need to look at the particular way you take tests. I'll discuss how you can evaluate that in chapter 9.

In a sentence: If you can't remember a fact, come back to the question in five or ten minutes. Don't assume that your first instinct *or* your second guess is more likely to be correct; trust your confidence in which answer is the right one.

TIP 52

Beware of "Pop Knowledge"

Some tests call for straightforward recitation of exactly what you memorized. For example, when a third grader takes a spelling quiz, she knows that the match between the quiz and what she's studied will be exact—she studied the spelling of a number of words and that's what will be tested. More often, test questions require some interpretation or application of what you've memorized.

The need for application opens the door to test takers choosing answers that aren't wrong but don't answer the question posed. For example, the instructor may ask the question "How did the Romantic movement in philosophy influence British poets of the period?" The student writes an essay that's crammed full of facts about Romantic philosophy and poetry but doesn't connect the two—which was what the question required.

Why would you fail to answer the question? Sometimes it's because you don't know the answer, of course; you just write whatever you know about the topic and hope for the best. But sometimes it's because you see a key term or two, think "I know this!," and start answering before you've read the question fully. I call this "pop knowledge." **When a question makes a certain piece of information pop into your mind, you must evaluate whether it actually answers the question.**

What amounts to the same thing can also happen in multiple-choice situations. For example, look at this question from a practice test for licensure as an electrician.

What must you observe when plugging a voltmeter into a DC circuit?

A. Power factor
B. Rms
C. Resistance
D. Polarity

The correct answer is "Polarity"—that is, which side of the circuit is positive and which is negative. That's what you have to observe, but a voltmeter is commonly used to measure resistance, and that's answer C. In studying for the exam, a future electrician will frequently encounter the directive "Use a voltmeter to measure resistance." So when the question includes "voltmeter" and one of the answers is "resistance" the test taker's brain immediately shouts, "Those go together!" And they do, but that's not the question. "Pop knowledge" may be factually accurate and appear frequently in study materials, but it can still be a bad fit to the question posed.

The best way to answer a test question is, of course, to read the question carefully. But in addition, you must be aware of your brain's tendency to serve up "pop knowledge" when you see key terms.

> *In a sentence:* When you've prepared well, some ideas will be strongly associated, and when you see idea A, idea B will immediately come to mind, but that doesn't mean idea B is the answer.

TIP 53

Ask the Instructor for Clarification, but Show What You Know

Sometimes you have trouble accessing something in memory because the instructor phrases things in an unexpected way. Part of the problem is that we tend to learn new ideas as expressed with particular phrasing, and if they are expressed differently we may not recognize them. That's one of the reasons I think a study group is helpful: studying with others exposes you to different ways of expressing the same idea (see tip 38).

Other times, the instructor has simply written a poorly phrased question. **If a question confuses you, you might ask the instructor about it.** Now, instructors vary a lot in their policy on this matter. Some instructors simply won't answer questions during an exam. Some college instructors aren't even in the room; they let teaching assistants proctor the exam while they do something else that's presumably more important.

Assuming that there is someone available and willing to field questions, you're more likely to get a helpful answer if you appreciate the instructor's perspective on the matter. Those of us who answer questions during tests are a little conflicted. On the one hand, we don't want you to get a question wrong because of some quirk in the way we've worded it. So we're happy to make sure that the question is clear. On the other hand, we don't want to give you a hint to the answer; that's not fair to everyone else. For that reason, we're sensitive to student "questions" during a test that seem like fishing expeditions: vague queries cast out in the hope that the instructor will unintentionally reveal something about the answer.

You're more likely to get a good response from the instructor if you offer reassurance that that's not your game. The way to do that is to **explain your confusion**. Don't just say, "I don't really get number four" or "Can you rephrase number four?" Instead, explain your confusion. Be brief, but also include specific details that show you've been thinking. Say, for example, "I'm confused by number four because it asks for an explanation of what's wrong with national education curricula, but we discussed several examples of national curricula where students do really well, like Hong Kong and Singapore and South Korea."

The instructor may draw your attention to a word or two that you had ignored. Or the instructor may decide that the question is not clearly written and offer a rephrasing that helps explain it. Or the instructor may tell you that you're on the right track and your confusion is just nerves. Or the instructor may not be helpful at all and just say, "Answer the question as best you can." But the instructor is almost guaranteed to say something like that if you don't let her know what you're thinking. If you just say, "I'm confused," she will think you're making a plea for in-exam help and she won't want to give it.

In a sentence: If you're confused by the phrasing of a question, ask the instructor to clarify, but be specific about your confusion and specify what you do understand as a way of reassuring the instructor that you're really looking for clarification of the question, not hints to the answer.

TIP 54

Don't Overthink

At the start of this chapter, I said that you use two key mental processes when taking exams: drawing information from your memory and then putting that information to use. We've looked at several ways that you can improve the odds of getting the right information out of your memory. What about putting it to use?

This is where test takers often go wrong, in particular on multiple-choice questions. When they are not 100 percent certain about the answer, they start applying what they think are clever test-taking strategies. These are actually methods of talking themselves into the wrong answer. Here are some examples of poor multiple-choice-question strategies and better ways of dealing with uncertainty.

Sometimes you just slip into overthinking. That's most common on multiple-choice exams because you are offered answers to which you can apply your overthinking. For example, you know A is right, and you know

B and C are definitely wrong. But then you look at D and think, "Hmmm. You know, D *could* be right." And without noticing you're doing it, you become an advocate for answer D—you try to think of circumstances that would make D a good answer. In so doing, **you will often add assumptions to the question and/or read things into it that are not there.** For example, one of my tests had a question about memory of an everyday event (going to a restaurant) and a student selected an answer that made sense only if you assumed that eating out was a highly emotional event. She said, "I thought you were trying to get at emotion and memory, so maybe eating out makes you really happy." After the fact, she saw that that reasoning didn't make sense.

If you find yourself unable to choose between two answers, ask yourself: (1) Do I need to add assumptions to make one of the answers correct? (2) Is one of the answers correct only under some circumstances? If the answer to either question is yes for one choice but not for the other, you have found the right answer.

Other times students don't slip into overthinking; they purposefully plunge into it. Students learn tricks to eliminate answers on multiple-choice questions when they take standardized tests such as the SAT— tricks like "Avoid answers that say something is 'always true' or 'never true'" or "If something is stated positively in one choice and negatively in another, the positive choice is usually right." **SAT tricks are your last resort.** They are what you try in desperation when you have exhausted every other option. Whether they work *at all* is open to debate, but even their advocates would say you are not meant to apply them to every question. That's how you talk yourself into wrong answers.

For multiple-choice items, try to answer each question mentally *before* you read the choices. If your answer is one of the choices, you're good to go. If the answer is not immediately obvious to you, the usual tricks suggest that you analyze the choices and start trying to eliminate them: Which one seems most likely to be wrong?

This is bad advice.

If you don't know the answer, you need to **spend more time on the question**. The answer must come from memory, and the question is your cue to memory. Work with the question to get to the answer. David Daniel, a learning expert at James Madison University, offers an "80/20 rule": most students spend 20 percent of their time on the question and 80 percent of their time thinking about the answer. They would be better off, he suggests, reversing that allocation and spending 80 percent of their time on the question.

In a sentence: Test-taking strategies that are supposed to guide you to the correct answer if you don't know the content don't work, and they often make you second-guess yourself.

TIP 55

For an Essay Question, Don't Start Writing Until You Know How the Essay Ends

Most students need to strategize less on multiple-choice questions but more on essay questions; they don't think enough before they begin writing their essays.

It's understandable. Essay questions are broad and often provocative, so the question immediately calls a few ideas to mind. Once you've jotted those on a piece of scratch paper, then brainstormed a little more, you might feel you can see the shape the essay will take. Given that you're under time pressure, you're impatient to start writing.

Screenwriters know this temptation well, as evidenced by an old saying. When one writer tells another she has a terrific script she is working on, the second will often ask, "How's your ending?" It's hard to start a screenplay, to invent vivid characters and an interesting situation for them. It's incomparably harder to end it in a way that will satisfy an audience.

The same is true of exam essays. If you've studied, you can probably think of a few big pieces of an answer, the parts that will enable you to write something that's okay. It's much harder to get the last parts into place and to organize them effectively. That's what will take you from okay to great.

I recommend a three-step process to writing essays. Suppose you encounter this test question: "Write an essay exploring 'appearance versus reality' as a key theme in *Hamlet*. Do you think Hamlet's inability to face reality is his undoing?"

Step 1: On scratch paper, list everything you think should be part of the essay. Go ahead and do a brain dump, but recognize that not everything you think of will be relevant to the question. There's a whole lot in *Hamlet* that has little to do with this theme, yet, because you know it, you'll be tempted to shoehorn it in.

To organize all the facts you list and to be sure that you include relevant facts in your answer, **sort them by subquestions**. In this example, the subquestions are (1) the exploration of the theme "appearance versus reality" and (2) Hamlet's inability to face reality as his key character flaw.

Organizing by subquestion will help you evaluate the evidence that you'll use to support your answers to these subquestions. You should have *at least* two sources of evidence for each subquestion, preferably three or four. If the claim is "A causes B" or "A is a type of B," there should be multiple reasons you say that.

Also, as you're doing your brain dump, think in terms of a hierarchy, just as I described for lectures and readings when you studied. You're listing facts, themes, ideas; they are not all at the same level of the hierarchy. A conclusion lives at the top of the hierarchy; underneath are amplifications of the conclusion, evidence for the conclusion, qualifications of the conclusion, and so on. Be explicit in your essay about what type of connections you are providing.

Finally, decide how much time you can reasonably be expected to spend on the question. More detail in each answer is obviously expected if there are three questions on an exam than if there are ten.

Step 2: On scratch paper, write an outline. From step 1 you have a bunch of facts, and you've started to connect them by considering which part of the question they are relevant to and the hierarchical organization of these facts. Now flesh that out into an outline. I know time is tight, but you want your essay to be well organized, and you want it to be composed of well-written sentences, but you can't think of two things—logical organization and clear sentences—at once. So write an outline that organizes your thoughts.

Think about the sequence of your ideas: How will you transition from one to the next? Outlining will also help you spot holes in your logic and places where you need more detail. It will also force you to think about your conclusion. **Don't start writing until you know how the essay will end.**

Step 3: Write. If you write an outline that you're happy with, you don't have to think about the answer to the question anymore—that's in your outline. Now you can focus on making your writing as clear as you can. You can think about word choice, making your paragraphs coherent, and varying the lengths of your sentences.

Instructors will sometimes say that only the quality of your thinking matters, not the quality of your writing. Maybe other instructors are better at grading than I am, because I find it really hard to separate the two. When I'm reading something and I'm confused, I have to judge whether the ideas are incoherent or the writing is poor. That's not always easy. Even if writing is said not to count, writing to the best of your ability sure won't hurt you.

> *In a sentence:* Essays require a lot of on-the-spot thinking, so use a three-step plan to put together an organized essay in a hurry.

With all the preparation you've done and the care with which you've taken the exam, you've maximized your chances of doing well. But of course things don't always work out. Maybe you started the course at a disadvantage because your background in the subject wasn't very good. Or maybe you simply got unlucky on the test. Or maybe your studying and test taking still need a little fine-tuning. In the next chapter we'll consider how to examine your exam results to figure out where to go from here.

For Instructors

It's not pleasant for instructors to contemplate, but we actually have no idea whether the tests we write do what we intend: faithfully measure students' skills and knowledge. Professionals who develop standardized tests spend a lot of time vetting individual questions for ambiguity and other flaws, but we're not professionals and we lack time. Still, some simple safeguards can help.

Pick the right question format. Multiple-choice questions are good for testing fine-grained distinctions among concepts. Fill-in-the-blank and short-answer questions are good for ensuring that students recall (not just recognize) simple ideas. Essays are good for testing students' ability to analyze and think critically. Don't kid yourself into believing that you can test critical thinking with multiple-choice questions. Some of the best constructors of tests in the country work on the National Assessment of Educational Progress; they've tried and failed.

Don't test students' ability to read instructions. Students will assume that your test should be taken the same way they've taken other tests. If you violate that assumption, make it even more clear to students than you think is necessary. For example, if you don't want a paragraph of prose in response to a short-answer question but instead want a list of reasons, make that very clear. If the word *not* is key to understanding the question, boldface and underline it.

Don't test students' luck or intuition. Tell the students what to expect on the exam. Do they need to memorize names? Dates? What percentage of questions will come from readings versus lectures?

Don't test knowledge unrelated to course content. Cute cultural references—for example, "Bart Simpson is weightless and moving at 50 km per minute"—are distracting for students who don't get them. Also, don't use complicated syntax, unnecessarily lengthy question stems, or multiple-choice options such as "A and B" or "None of the above." These test students' ability to keep a lot of information in mind at once.

Don't test students' skill in interpreting ambiguous questions. Because you can't write perfect questions, you need some way that students can get clarifying information during the test. Either make yourself available during exams or allow students to write marginal notes explaining why they answered as they did.

Much of what I've advised here may strike you as hand-holding, as providing perhaps too much help to ensure that students understand what's being asked of them on an exam. I think it's more accurate to view it as ways of ensuring that your test measures what you intend it to measure.

Summary for Instructors

- Match the question format to the type of knowledge to be tested.

- If you violate students' expectations about the typical ground rules for exams, make that very clear.
- Tell students beforehand what content they are expected to know for the exam.
- Don't include cultural references or, more generally, extraneous information in questions.

CHAPTER 9

How to Learn from Past Exams

Suppose you take a test and it goes poorly. Clearly something about the way you prepare must change, but what? Most people conclude, "I need to study more." That isn't helpful because it isn't specific.

Consider all the reasons you might have missed a given question:

1. You were never exposed to the content it tested because you were absent for that lecture or skipped the relevant reading.
2. You were exposed to the content but didn't understand it.
3. You understood it, but the content didn't make it into your notes.
4. It was in your notes but not your study guide.
5. It was in your study guide, but you didn't memorize it.
6. You memorized it but were momentarily unable to recall it during the test.
7. You would have been able to recall it from memory, but you misinterpreted the question.
8. You had the right answer in mind, but you accidentally circled the wrong choice on your exam paper.

It's likely that you make some of these mistakes frequently and others infrequently. It's not fun to dissect a failed exam, but you need to diagnose your area of greatest need so you know where to put your effort going forward.

WHEN EVALUATING WHAT WENT WRONG ON A TEST

What your brain will do: It will make a snap diagnosis about why you failed: "I needed to study more."

How to outsmart your brain: Overcome the impulse to turn away from failed work, and analyze what went wrong. That analysis can direct your effort for the next exam.

The tips in this chapter tell you how to use completed exams to identify problems in your preparation. It will also address a couple of common problems students encounter when doing this analysis.

TIP 56

Categorize Your Mistakes

Figuring out what went wrong on an exam means analyzing the questions you couldn't answer. Start by flagging those, but **also flag the ones on which you guessed and got lucky.** You couldn't answer those, either.

Now, how should you analyze them? Here I'll cover exam questions for which the test writer had a specific answer in mind, such as multiple-choice questions, fill-in-the-blank questions, and calculation problems you would find on math or science exams. There are two ways to evaluate your mistakes.

First, you can **analyze the content** of the questions you got wrong. The most obvious way to do that is by subject matter. Did you miss a lot of questions that were based on a particular topic? Did you miss mostly questions based on readings or based on lectures? Did the questions you missed concern facts and details or more big-picture themes? If you can identify a pattern in the content of the questions you missed, you should pay attention to the completeness of your notes and study guide before the next test. Check with your study group to be sure that you've got all the content down.

Check to see whether the content you missed was in both your notes and your study guide. If not, you're skimping on your study guide. Next time, make sure it has *everything* in it.

Did the questions you missed demand straightforward recall of concrete information, or were you asked to use the knowledge in a new way? Application is always more difficult, but you can get better at answering such questions (see chapter 6).

Second, **analyze what went through your head when you saw each question you got wrong on the test.** Here are eight common thoughts people have when they review questions they got wrong, along with what each thought probably means.

1. **I was surprised that the question was on the test.** This means you either missed the content altogether (that is, it's not in your notes) or you judged that it was unimportant and so didn't put it into your study guide. Missing a question or two for this reason is pretty typical, but if you missed several

questions for this reason, the remedy is obvious: you need to be more careful to make sure your study guide is complete.

2. **None of the answers looked right to me (in a multiple-choice question).** Possibly you understood the concept but failed to include it in your study guide, or possibly you don't understand the concept although you think you do. The most likely scenario, however, is that what's in your notes or study guide is not quite right. Comparing your understanding of the material with other people's can help (see tip 23).

3. **The answer seems clear enough to me now, but I couldn't recall it at the time.** You didn't study your study guide enough. You probably needed to do some overlearning (see tip 50). You can also review the tips on memory recall in chapter 8.

4. **I'm told that this question tested a particular concept and I studied that concept, but I didn't see how it related at the time.** I've mentioned straightforward recall of information versus application of ideas; not seeing that a concept is relevant is an application problem. You may study how Pavlov's dog learned to salivate on hearing a bell and study that type of learning in a few other contexts (e.g., a child is scratched by a cat and then fears cats) but then not see that a situation described in a test question (e.g., coming to like a perfume because an attractive woman wears it) is the same type of learning. These are some of the most difficult questions, and chapter 6 describes how to prepare for them.

5. **I made a stupid mistake.** You started to read the question, recognized some key terms, and were sure you knew what it was about, so you wrote your short answer—but you didn't notice the word *not* in the question. Or on a math test where you were meant to apply $(x + y)^2$, you forgot to square. These mistakes are surely the most frustrating, but fortunately they

don't reflect a deep problem. You just need to take more seriously the advice to check your work noted in chapter 8.

6. **I still don't see why my answer is wrong.** Most likely, your notes and/or your study guide does not have fine enough detail in it. There's a concept that you partially understand, but you are missing an important detail that is preventing you from seeing why your answer is slightly off the mark. Check with the instructor to get further information.

7. **I overthought.** Overthinking arises when you employ a test-taking strategy. Either you talked yourself into a wrong answer or you talked yourself into a strange interpretation of what the question asked (see tip 54).

8. **It was a trick question.** You feel you knew the content and if a straightforward question about it had been posed, you would have gotten it right. Instead, the question led you down the wrong mental path because the phrasing was misleading. We'll take up trick questions later in this chapter.

If your mistakes tend to fall into one or two categories, great; you have a good idea of what to work on. Read the tips in the relevant chapters of this book and see if following them helps on the next exam.

If your analysis indicates that there's not just one or two problems—that is, you're getting lots of problems wrong for lots of reasons—it may be that your root problems are planning and organization. We consider those challenges in chapter 10.

In a sentence: Analyze the reasons you got questions wrong by considering what you were thinking when you tried to answer them; that will tell you which step went wrong when you prepared for and then took the exam.

TIP 57

Analyze What Went Wrong on Essay Questions

You would think it would be easy to analyze what went wrong on exam essay questions. Multiple-choice, true/false, or fill-in-the-blank questions provide minimal feedback, but for essays you expect more detail. Of course, you don't always get it. Providing this sort of feedback is time-consuming for the instructor. I may, when I write an exam, have every good intention of providing clear, detailed feedback on each essay, but when confronted by eighty exams and limited time, I end up writing useless comments such as "Vague" next to a big block of text. (I still remember a comment my professor wrote on an essay for my final exam of a twentieth-century American literature course. The professor's feedback, in its entirety was "No. C+.")

If you get minimal feedback, you can always seek more information from the instructor. If that's not an option, at least see if someone else in your study group got full credit for questions you found difficult. Seeing what was judged to be a good answer may help you identify what your answer lacked. For example, you may see that your classmate provided more detailed examples or pulled together evidence from more course topics. Then you can work backward to figure out how to prepare more effectively. (For some standardized tests—for example, Advanced Placement exams—you can view sample essays with explanations of how they would be scored.)

You should also **consider what sorts of essays you were asked to write.** Two types of questions dominate essay exams. Some ask you to elaborate on and explain content. Perhaps you spent a full day in your microeconomics class discussing elasticity of economic variables. An essay question might be "Define *elasticity* and name three ways it can be measured, along with the advantages and disadvantages of each." This question demands direct recall of content—content that ought to be in your study guide and memorized. It is also, by the way, an easy type of question to grade. The grader knows exactly what she's looking for and can set a point value for each expected part of the answer (a definition, three measures, and the advantages and disadvantages of each measure).

It's therefore easy for you to assess what went wrong if you didn't get full credit. Just as you did for multiple-choice and short-answer questions, you need to size up whether the content didn't make it into your notes, didn't make it into your study guide, never got memorized, and so on.

In the second type of essay question, you're asked to evaluate something new: a conclusion, perhaps, or a hypothetical situation. There are a few ways your answer to this sort of question can go wrong.

First, **the instructor might have a particular answer in mind, and you just don't see it.** Someone preparing to be a teacher might take a class on reading instruction, and the final exam poses this question: "Would it be a good or bad idea to offer an eight-year-old a dollar for every book she reads over summer vacation?" You can't think of anything you've studied that addresses this question, so you just write what you hope is a coherent answer, tacking on ideas that seem relevant as they come to mind. Because the question didn't include the words *reward* or *motivation*, you have forgotten that half of a lecture was devoted to the relationship of reward to motivation—specifically, the idea that rewarding people to do something can actually backfire and make them less motivated to do the rewarded activity.

This problem is similar to one described in chapter 6: the question requires that you see beyond the particular circumstances (reading, money) to the underlying principles (motivation, reward). If you're expecting deep questions like this on the next exam, see the tips offered in chapter 6.

The second way this sort of essay can go wrong: you may be on the right mental track, but you end up writing a poor essay because it doesn't make an argument, it's unorganized, or you don't use transitions well, so the instructor can't see how the whole thing hangs together. **You've got a lot of the right facts in your essay, but you don't put them together so that they build something larger.** The example question about paying an eight-year-old explicitly demands that you draw a conclusion about the idea. You should probably point out both advantages and disadvantages of the payment, but at the end you need to weigh the evidence and conclude either "good idea" or "bad idea." If that conclusion is missing or doesn't seem justified, your essay could have been better.

A third possibility is that you do remember the relevant content and you include it, but **you clutter your essay with a bunch of irrelevant stuff**. You're pretty sure you're supposed to discuss rewards and motivation, but you think there's always the possibility that the instructor had something else in mind, so it seems like it can't hurt to add other stuff to your essay. So you write about why reading is important to success in school, you summarize things you remember from a developmental psychology class about what eight-year-olds are like, and you discuss how behaviorist psychologists use rewards in their theories.

Students often figure, "The more I show what I know, the better." Maybe, but usually not. Some instructors specifically tell you that they will lower your grade if you load your essay with irrelevancies because you are hoping for points. Even if it's not a policy, when I'm grading, it's hard for me to overlook that you've got three good points in your essay and four facts that are true, but none is related to the question. It's like someone serving me ice cream with beef gravy.

"What's the matter? You don't like gravy?"

"I do, but not here."

The way to avoid this problem is to be more critical about what to include in the essay when you write your brief outline.

A fourth possibility is that **your essay is good, but it's a mediocre fit to what the question asked**. For example, suppose a final exam question in a Shakespeare course asks you to compare the playwright's view of love in Shakespeare's plays and in Greek drama. The instructor expects you to focus on *Romeo and Juliet*, but for some reason you barely mention that play and build your answer on the other Shakespeare play you read, *Hamlet*. It's not a bad essay, but you were on the wrong mental path from the start. What you needed to do was brainstorm longer before you started to write, even before you started to outline. You probably thought of *Hamlet* first and in nervousness lunged toward that answer.

Finally, let's consider poor writing. Most graders will not deduct points for grammatical errors, misspellings, errors of usage, and the like. They *might* deduct a point or two if you adopt a very informal, inappropriate voice in your answer, for example, "People think Kant is deep and whatnot, but a lot of times when you read him, he just seems nuts."

Now, "no penalty for grammar" might be the policy, but if your essay is full of grammar mistakes and you're on the borderline of two grades, a grader might unconsciously fail to give you the benefit of the doubt. That said, if you get a poor grade on an essay, don't conclude, "I guess she just doesn't like my writing." Teachers are experienced and are accustomed to many different writing styles. Put a little more time into a final proofread at the end of the exam.

In a sentence: Even if the grader provides very little feedback regarding why you earned the grade you did on an essay question, if you know the typical ways that essay questions go wrong, you can figure out why you scored poorly, and you'll know how to improve next time.

TIP 58

See Trick Questions for What They Are

Here's a riddle: Imagine you're in a rowboat that's sinking. There's no land in sight, and you're surrounded by hungry sharks. What should you do?

Answer: Stop imagining.

Why do people groan when they hear riddles like this one? Because they expect the answer to require some cleverness, some ability to solve problems. Instead, getting it right requires that they assume bad faith on the part of the riddle teller—bad faith because when I tell you a riddle, I'm inviting you to imagine, to pretend that this invented world is real and that things behave in the riddle world the way they do in the real world. Without that rule, riddles are pointless; when you pose the sinking-rowboat riddle, I could simply say, "I pull a helicopter out of my pocket and fly away."

Trick questions on exams resemble a bad-faith riddle. Someone who knows the content would answer the trick question *this* way, but the instructor has a devious justification for another answer. The student sees "2 + 3 = ?" and writes "5," only to be told, "No, no, that wasn't a plus sign, it was a rotated multiplication sign. The right answer is 6."

I think that **trick questions on exams are actually quite rare**. People who write tests want to find out what test takers know. If they are instructors, they also want students to enjoy the course and to appreciate the subject matter. Both goals are undercut by trick questions.

If you think your instructor poses questions that call for a lot of subtle interpretation, check with the other people in your study group. I'll bet they think the question you thought was tricky is actually pretty clear—but each of them had a different question or two that *they* saw as tricky.

When a question looks tricky, **the problem usually lies in the student's knowledge of the content, not in the wording of the question**. For example, suppose you see this question:

Paintings from the Romantic era in Western Europe:

A. focused on landscapes and seldom included human figures.
B. often showed forces of nature at work and included human figures.
C. focused on themes from Greek mythology.
D. were wholly religious.

You know that the Romantics did not like the Classical period and weren't religious in the traditional sense, so C and D aren't right. You also know that the Romantics focused on nature, but you have a hard time choosing between A and B. You finally settle on A because it seems like there being no humans means that there's a greater emphasis on nature. But the right answer turns out to be B. You're irritated because the two answers seem so close and the right answer seems inconsistent with what you thought you understood about the Romantic era. The main difference between answers A and B seems to be whether human figures are seldom included or included. So it all seems to come down to the definition of *seldom*, which seems really subjective.

But your interpretation wasn't quite right, because your knowledge of the content wasn't quite deep enough. You knew that Romantic painters sought to portray nature, but you didn't know that they especially focused on its awe-inspiring power. People appeared in their paintings as spectators to the splendor of nature; it's important that the figures were often tiny, because that highlighted their insignificance.

Questions can also seem tricky because "pop knowledge" doesn't work. For example, suppose the instructor used "The curtain of night fell" as an example of a metaphor. Most students have that example in their notes and study it. Then the exam contains this item:

"Night fell like a gentle snow" is an example of:

A. a simile.
B. a metaphor.
C. an analogy.
D. none of the above.

Having studied "The curtain of night fell" as an example of metaphor means that your memory will have those concepts bundled together. When you read "Night fell" in the test question, *metaphor* pops into your mind. But of course the use of the word *like* means it's a simile, not a metaphor. I avoid writing questions for which "pop knowledge" leads you to the wrong answer, but you will see them on exams (see tip 52).

It's true that sometimes instructors inadvertently include multiple-choice questions with two answers that are defensible or with a question for which the phrasing is confusing. Good instructors will admit that's the case and give credit for both answers. But don't assume that's happened if your answer seems right to you but is marked wrong. What's more likely true is that you understood the content well enough to choose an answer that's close but not deeply enough to pick the right one.

In a sentence: Most of the time that you find a question tricky or confusing, it's because your knowledge of the content isn't quite deep enough.

TIP 59

Think About What Went Right

The fact that you seek out information about what went wrong on a test shouldn't mean that you don't acknowledge and appreciate what went right. You learned something, even if you didn't do your best and even if you're disappointed with your grade. **Give yourself credit for the work accomplished.** People who are depressed and hopeless about their work are (1) not being realistic, (2) likely to be less motivated for future work, and (3) really boring company. Snap out of it.

But this is not only about mood and motivation. You analyzed the questions you missed to figure out what not to do. **You should also analyze the questions you got right, to figure out what you should keep doing.** Are you nailing the details? Are you great with the readings? Are you shrewd about not getting suckered by "pop knowledge"? Whatever it is, take a bow and keep it up, especially if your success is the result of trying something new in your studying.

Analyzing the questions you got right may also **refine your sense of what you need to work on**. For example, when you examined the questions you got wrong, you may have noted that a lot of them asked you to integrate ideas from different lectures. But then when you look at the ones you got right, you see that there are a number of cross-class integration questions among them. So now you can ask, "Is there anything different about the ones I got right and the ones I missed?" Maybe you'll see that you were pretty good on those questions early in the marking period, but as things got busy, you no longer had time to integrate ideas when you reorganized your notes. Assessing your strengths may help sharpen your understanding of your weaknesses.

Knowing what to change in your future work requires acknowledging both hits and misses.

In a sentence: Pay attention to what you got right, both because it will make you feel more encouraged and because it will help you refine your understanding of what you need to work on.

TIP 60

Don't Cringe

Every now and then I find myself in a conversation where someone describes what their personal experience of Hell would be. For years, I volunteered that my vision would be the Devil ushering me into a small room containing nothing but a low stool, on which I would sit while someone read my college senior thesis aloud to me for eternity. So if you're depressed by the prospect of performing an autopsy on a failed exam, I understand.

Norman Vincent Peale said that most of us would rather be "ruined by praise than saved by criticism." Still, you *can* overcome your reluctance to go over what went wrong on an exam.

Some people draw the wrong conclusion about what a failed exam says about them because they have a warped perspective on schooling and intelligence. They believe that:

1. You're born either intelligent or unintelligent, and that can't be changed.
2. Intelligent people don't make errors.

If these statements are true, that implies that if you make errors, you're showing the world that you're unintelligent, and nothing can be done to change that because intelligence is inborn. You can see why poring over errors would be pretty threatening to your sense of self. Fortunately, these premises aren't true.

Let's look at what research says about the changeability of intelligence. Intelligence has two components: how much stuff you know and the ease and speed with which you can move information around in your mind. That second factor—what we might call "mental speed"—probably can't be changed. People have tried to develop training programs to improve it, but no one has succeeded, at least, so far.

But the other factor—what you know—is easily changed. **Learning more information makes you more intelligent.** Learning can be discouraging, though, because the people who are good at the mental-speed part are better than the rest of us when we take on a new task. In other words, if two people learn how to play chess, the person with good mental speed will pick up the game faster and will beat the person with slower mental speed. *But* if the second person practices, she'll gain chess knowledge—of standard openings, for example—and she'll soon defeat her high-mental-speed opponent who lacks that knowledge.

You can get smarter in any subject you want to. You just need to learn the subject.

The second premise—that smart people don't make errors—is also obviously false. Who doesn't make mistakes? It's probably true that the people you think of as smart don't make as many mistakes, but that's because they work hard.

People at school sometimes like to claim that they didn't do the reading, didn't study for tests, and so on. They make that claim because it fits the belief that smart people are just naturally smart and don't need to try hard. I've been in schools as a student or an instructor literally my whole life, and I can confidently say that the students who do well in school work hard, with extremely rare exceptions.

Part of working hard in school is figuring out what you don't do well so you can focus your energy where it's needed. The person who gets all As is the person who is not afraid to learn from their mistakes. **Going over exam mistakes may make you *feel* dumb, but you're actually doing what smart people do.** You should remind yourself of that.

There are other things you should say to yourself as well. Remind yourself of how far you've come. Okay, maybe you underestimated what was needed to reach your goal, but that doesn't negate what you've achieved. What would have happened if you hadn't worked as hard as you did? You also might want to remind yourself of why passing this course, certification exam, or whatever matters to you. Your dream was not to get an A on this quiz; you have a larger, long-term goal. You still have that goal, and one setback should not dissuade you from continuing to work toward it.

If facing your failed exam still feels like a terrible burden, here's a method that might help you the first few times.

To get over your initial reluctance, promise yourself that as a first step *all* you will do is categorize your errors (see tip 56). You won't race to look up the right answer or mentally defend the answer you gave. You'll just sort the questions you missed. If you start to beat yourself up about missing questions, you will say aloud, "I'm doing what smart people do after an exam. This feels lousy, but it's the right thing to do." Sometime later, go back and look at the material in the readings and your notes (as needed) to get a better sense of exactly what happened on each question you got wrong; you may change your mind about which category some belong in. Sometime after that, evaluate the consistency of the mistakes you made. Splitting the work up into different sessions may seem counterintuitive, because you're eager to get it over with. But breaking the job down into smaller pieces will make it feel less threatening.

> *In a sentence:* You may think that successful learners don't make many mistakes; they do, but what separates them from unsuccessful learners is their willingness to face their mistakes and learn from them.

For Instructors

I think it's valuable to go over a graded exam in class but not to explain why one answer is right and another wrong. I think you should help students analyze the types of errors they made, as I did in tips 56 and 57. This sort of analysis will be unfamiliar to many students, and it's a tool they can use in other classes.

Students still need some way to get explanations about why answers are right or wrong, of course. That mechanism might be individual meetings, opportunities to contact you online, or an annotated answer key. I like to meet individually with students who want feedback about exam answers, because it's an opportunity for me to have a deeper conversation about obstacles to their learning and to talk about study techniques, note taking, and all the rest.

An aspect of these meetings that some instructors—especially in higher education—don't relish is that they can be emotional. The students desiring feedback are usually the ones who are failing. In fact, they often want to see you less for abstract advice about learning and more because an academic disaster is imminent. These students are upset.

In fact, sometimes a desire to meet with you is more about emotion than anything else. The student wants to be heard. He is mostly mad at himself and doesn't expect you to do anything. He just wants you to know that he is disappointed in his performance.

Other times the test is an excuse. The student is reaching out because she has a serious life issue. I teach at a university where most students are from comfortable families, but in the last few years I've had in my classes (1) a student whose parents had cut him off and was working long nights as a bartender to earn enough money to stay in school; (2) a student who had had to take in his niece because his sister had developed a drug problem; (3) a student who was living in the Charlottesville bus station; and (4) many, many students with nascent or fully developed depression or anxiety.

Sometimes a low exam score serves as a trigger for a student with difficult life situations to see you. But sometimes, too, such students seek help with their studies but don't reveal anything about their circumstances. Educators must keep their eyes and ears (and hearts) open to discern why a student has sought them out.

And when a student seeks you out, do not underestimate the power of your words. If you have taught for a while, you have probably had an old student return and recount in detail a conversation you may not even remember but that proved highly important to the student. We love these stories because we are the hero. I sometimes wonder how many times I've said something negative or impatient and so created a moment that was memorable for the wrong reason. Keep in mind how vulnerable many of your students are.

Summary for Instructors

- Use class time to model a test autopsy.
- Offer an alternative mechanism by which students can get

details about the factual content of questions and answers—
that is, why particular answers are right or wrong.

- Meeting one-on-one with students about their exam performance is time-consuming but is an effective way to have deep conversations about obstacles to their learning.
- Remember that students may be struggling in your course because they are experiencing serious life issues that they are reluctant to share.
- Remember that your words carry more power with students than you may realize.

CHAPTER 10

How to Plan Your Work

This chapter is written for people who have never gotten into the habit of planning their work. Surprisingly, that includes most college students. When they're surveyed, the most common answer to the question "When you're studying, how do you choose what to work on?" is "I do whatever is due next."

We'll take up two aspects of planning: remembering to carry out tasks at the right time and making sure you have enough time to complete them. Remembering to do things calls on **prospective memory**; that's what it's called when you form an intention to do something in the future and then later remember to do it. It's the type of memory you rely on when you notice that you're low on gas in the morning and think, "I should buy gas on the way home tonight." Another example of prospective memory is taking medications: you pick up your pills at the pharmacy knowing that you must remember to take one pill three times each day for the next five days.

Prospective memory can fail, of course—you forget to buy gas or take your pill when you planned—but the solution seems obvious: don't rely on memory. Instead, **set up a reminder to prompt the action at the right**

time: you might leave a note on your steering wheel so that you'll see it when you start your drive home or set an alarm on your phone for the time to take your medication. It's a good strategy, but setting up something to remind you needs to be consistent if you're going to rely on such reminders.

The second aspect of planning is judging how long the activity will take to complete. **People consistently underestimate how long it will take to get things done.** This is called the "planning fallacy." Just think about the last time you read about a major public construction job; they seem to run late and over budget every time. For example, the Sydney Opera House was supposed to open in 1963 at a cost of $7 million (Australian) but was finished ten years late and cost $102 million.

Project planners aren't stupid, but they are overconfident that their solutions to difficult problems are likely to work. For example, one obstacle in building the Sydney Opera House was the failure of the system designed to divert stormwater. In addition, people tend to completely disregard a potential problem if they think it is unlikely to occur. That sounds smart—why worry about something that probably won't happen? The catch is that there are *lots and lots* of ways a complicated project can be delayed. Each is, on its own, very improbable, so we ignore all of them, but taken together, one of the problems is actually pretty likely to happen.

The fallacy is easy to address: once you accept that it's real, you just need to allocate more time for work than you think you'll need. Guarding against prospective memory failures is trickier. The remedy is easy enough to describe: you need to make a habit of writing down what you're supposed to do *and* a habit of checking your to-do list. But developing the habits takes some perseverance.

WHEN PLANNING YOUR WORK

What your brain will do: It will not allocate enough time to complete scheduled work, and it will forget that you've planned that work.

How to outsmart your brain: Establish a small set of simple habits to make sure you know what work you're expected to complete and by when.

In this chapter I'll suggest something of an end run around the problem of planning. Scheduling becomes greatly simplified when, instead of planning time to work on each project, you **plan to work for a consistent amount of time each day.**

TIP 61

Get Enough Sleep

People—especially students—tend to treat sleep as an optional activity, making the odd assumption that it will somehow take care of itself or that they will catch up on sleep during the weekend. It's common for people to experiment with sleep in ways they would not experiment with other basic needs, like food or breathing.

Yet sleep has a direct effect on your cognitive performance. It's easy to appreciate that **sleep loss makes it both harder to think and harder to**

pay attention the next day. It also makes people's mood more erratic, so they're not that much fun to be around. What's more surprising is that sleep loss also messes up learning from the *previous* day. What you learn today goes into memory today, but there's another process by which the memory "gels," becoming more final and more resistant to loss. That process depends on sleep. **Thus, losing sleep disrupts what you learned the previous day.**

According to the US Centers for Disease Control and Prevention, teenagers should get between eight and ten hours of sleep each night and adults seven to nine hours. The estimated percentage of people actually getting that much sleep varies from study to study, but it's likely less than 50 percent.

Most of us don't get enough sleep because we get to bed too late; that is, the problem is not that we pop awake early in the morning, before the alarm rings. Although sometimes you are kept awake later than you wish, oftentimes you just don't feel sleepy when it's time to go to bed. Why?

Your body is sensitive to two cues that signal "It's time to sleep." One is the body's internal clock, especially the production of a hormone called *cortisol*. Cortisol is like an alarm. Your body produces lots of cortisol in the morning and less in the evening. Your body's internal clock is most noticeable when it becomes disengaged from the time of day; when a Londoner travels to Toronto, she may get sleepy at 6:00 p.m. because her body thinks it's 11:00 p.m. During the teenage years, the peaks and troughs of cortisol production flatten somewhat, which is one reason teens aren't sleepy at night and have a hard time waking in the morning.

Your body also pays attention to signals out in the world. For example, if you have a nighttime routine—you brush your teeth, wash your face, put on your jammies, dim most of the lights, and read for a few minutes—your body learns your routine, and after you do the first five things, it knows that it's time to sleep.

The research I've just mentioned points to concrete steps you can take to get better sleep. The time you wake up in the morning is probably out of your control, so sleeping more means getting to sleep earlier. You can change the external cues relatively easily. Your internal cues will lag behind, but they will eventually adjust to the external cues; that's what happens when you recover from jet lag. Here are some methods of changing the external cues.

1. **Have a consistent routine.** You may feel a bit silly creating a before-bed ritual, and it will take time for your body to learn it, but it will help you go to sleep more quickly. Part of that routine is a consistent time to go to sleep. With practice, your internal clock will attune itself to it so your body knows when to make you sleepy.

2. **Avoid looking at screens for an hour or two before you sleep.** The light from a screen is a cue to your brain that it's closer to the middle of the day than it really is, so it confuses your internal clock. When someone has a hard time sleeping, they often turn to their phone, figuring, "I'm on my phone because I can't sleep." But the reverse may be true: they can't sleep because they are on their phone.

3. **Just lie there.** I know that sounds weird, but you don't want to send your body mixed messages. When you pick a time to go to sleep, stick with it; don't lie there for five minutes, conclude that it's pointless, and get up. Just lie quietly with your eyes closed and figure that you are at least resting.

4. **That said, use common sense when you pick your target sleep time.** Suppose you have been getting to sleep around 2:00 a.m. each night and you would like to start going to sleep at 11:00 p.m. Don't get into bed at 10:59 and just lie there. Shoot for a thirty- or even fifteen-minute backward increment. Make sure you're in bed by 1:45 a.m. for a week or how-

ever long it takes until you regularly fall asleep pretty quickly at that hour. Then make it 1:30 a.m., and so on.

5. **If you can, nap during the day.** Some people have enormous difficulty napping, I know. Their bodies just won't cooperate. But others can, and it's a good way to achieve more sleep time if you find that you often have interesting things you want to do late in the evening. If you find you're extremely groggy when you wake from a nap, that's a sign that you've been deeply asleep. Try napping for no more than twenty minutes, and do so in a not-terribly-comfortable position—for example, in an easy chair. That might keep you from falling too deeply asleep.

In a sentence: Sleep directly affects learning, and although many people are frustrated by their inability to sleep as long as they want to, there are steps you can take to help you get more sleep.

TIP 62

Plan a Block of Consistent, Dedicated Time for Learning

When a student sits down to work, she is typically not strategic in deciding what to do; she works on whatever happens to be due next. This strategy can scarcely be called planning; it's damage control, and it leads to three undesirable outcomes.

First, it leads to cramming. If Tuesday is filled by studying for things due Wednesday, and Wednesday is filled by studying for things due Thursday, you can't start studying for Friday's quiz until Thursday night.

Second, if you're in the habit of asking yourself, "What's due tomorrow?" on days when the answer is "Nothing," it's natural to think, "That means I have the day off." You end up studying less than you would, in a more reflective moment, say you ought to.

Third, counting on external deadlines (such as tests) as your guides to studying becomes a habit that's hard to break once you're out of school. You will still look to deadlines at work for motivation. Say you intend to teach yourself to code because it would help your long-term career prospects. A lifetime of being prompted to study only when a test looms means that you're unlikely to make time for learning when there's no urgency.

A better strategy is to **plan your learning by time, not by task**. In other words, plan a block of time each day that is dedicated to learning. If nothing is due tomorrow or even for the next several days, work on assignments that are due later.

Treat this block of time as unchangeable; don't schedule appointments at that time or skip a day if something you deem more important comes up. Think of it as a job for which you must show up. Accordingly, schedule it at a time you know you will be able to honor.

Planning by time rather than task brings important advantages:

- Your memory for whatever you study will be much better if your studying is spread across days. That's a product of the positive effect of sleep that we just discussed and of the spacing effect (see tip 39). You'll actually get more done in the same amount of time if you spread the work out.
- Spacing your work out gives you more flexibility if you misjudge how long it will take to do something or if something

that *should not happen* nevertheless does—for example, your roommate locks you out of your room and goes home for a few days. If you wait until the night before something is due, the extra time you need or the unexpected obstacle will pose a real problem. But if the due date is still a few days away, you can adjust.

One safeguard against the planning fallacy is being willing to work on assignments well in advance of their due dates if your other work is completed. Obviously this method works only if you set aside a sufficient block of time for learning each day, but fortunately there's a built-in mechanism to let you know if you are allotting enough time. Suppose you see on your calendar that you have a math quiz in two days. You've done nothing to prepare for that quiz today, and your designated work time is over. What should you do?

You should do some studying for the quiz even though your work time is supposed to be over. More important, you should increase your daily study time, perhaps by fifteen or even thirty minutes. Yes, even though it was inadequate just this once. **You will never regret being ahead on your work. This is your insurance policy against the planning fallacy.** Rest assured, fate will snatch away your head start by making your computer fail or creating some other unexpected obstacle.

All right, you've adopted a routine of working approximately the same amount of time each day at around the same time of day. How should you decide what to work on?

In a sentence: Instead of planning your work assignment by assignment, make a habit of working a set number of hours each day.

TIP 63

Use a Calendar

I've suggested that you plan your learning by time, not by task; dedicate a set amount of time each day to learning. Thus, you sit at your desk at 7:30 p.m., for example, ready to work for two hours, and you ask yourself, "What learning tasks am I taking on today?"

To answer that question, you need to know what tasks have been assigned to you (or that you've set for yourself) and when they ought to be finished. You need that information written in one place so you can see everything you need to do simultaneously, especially because there's so much to do: reading, reorganizing notes, preparing for exams, and so on.

If you search the internet for information about how to use a planner, you will find guidelines written by the sorts of people who really enjoy using a planner. Their videos show you how to use four different highlighter colors to make different types of activities stand out. They show how to draw arrows and borders for emphasis and how to draw shadows on capital letters, and they suggest that you maintain multiple lists (daily to-do list, weekly to-do list, monthly planning, birthdays, reading journal, reference, gift ideas, shopping, and so on). If you enjoy planning, this kind of thing is satisfying.

My father was like that, so I've observed the benefits of exacting planning and organization. Nevertheless, that's not me. My natural inclination is toward disorder and chaos. Until graduate school, my time management system was a mix of writing things on my hand, apologies, and excuses.

The fast pace of graduate school forced me to become more systematic, and I found that using a very simple calendar provided a huge benefit in productivity compared to trying to keep all my responsibilities in my head. I was so pleased that I experimented with making my simple system a little more sophisticated. But the increased payoff was small, and I soon returned to my elementary calendar.

There are two mandatory principles of using a calendar. You must do these two things, but if you do them, you don't need to do much else.

Principle 1: Have your calendar with you all the time. You don't know when you will need to schedule an assignment, social event, airport pickup, or whatever, so it needs to be with you. Hence, keeping a calendar on your phone is a good choice because you're already in the habit of carrying it. Some people prefer a paper calendar because of the larger format or because there's something they like about the feel of paper. Some students have classes that ban electronic devices, so phone-based calendars are not a great option. Use either one, so long as you make a habit of carrying it. If you want to carry a paper calendar but you always forget to have it with you, set it by the front door in the evening (see tip 6).

Principle 2: Write commitments in your calendar immediately. You can't trust that you'll remember something later, so you need to record it the moment you learn about it. If you find that hard to remember, try setting an alarm on your phone for two minutes after each class ends; that will remind you to consider whether there were any new assignments that should go into your calendar, so at least your class assignments will be covered.

You can get away with just one calendar (as opposed to adding a weekly and/or monthly calendar) by *immediately* adding reminders to your daily calendar for commitments that require preparation for more than one day. In other words, if an instructor announces a math quiz on September 28, put that into your calendar and also add "Math quiz in five days" on September 23 and "Math quiz in three days" on September 25. **These early reminders of upcoming deadlines are crucial; they will help you avoid the planning fallacy.**

Note that you don't write these notes on a separate to-do list; you write them on the calendar entry for when the class meets. Separating assignments from the dates they are due makes no sense. When you sit down to work, you must decide "What is my highest priority?"; the due date of each assignment is obviously a crucial factor in that calculation. So why separate what needs doing from when it's due?

If you get assignments in bulk at the start of the marking period, add them to your calendar as soon as you get them. Add calendar entries for assigned readings and for lecture notes that you will reorganize (see chapter 4). Get your school calendar and note vacation days, important school events, and so on. Block out the time you've chosen to study each day (see tip 62).

Don't neglect to **put scheduled social events onto your calendar**. That doesn't apply just to formal things like parties or arrangements to meet with friends. Block out time if there's a football game you know you will want to see or if your favorite artist is releasing music and you know you'll want to download it and listen to it immediately. You need to be able to see all the claims on your time.

If you can develop these two habits—always have your calendar with you, and always record new obligations as they come up—you will know what you are supposed to do and when it's supposed to be done. So the problem we started with—sitting down to work and deciding what to do—will be mostly solved. But we can elaborate just a bit more on how to go from what's recorded in your calendar to how to plan your day.

> *In a sentence:* If you're not already using a calendar, you must start; it's essential to managing your time and setting priorities for learning.

TIP 64

Make a To-Do List for Each Study Session

You might find it useful to create a daily to-do list for all your activities. I've never used one, but then, I've already told you I'm not all that organized. I remember to do laundry because I notice I'm out of socks, not because it's on my to-do list.

But. For each study session you must make a to-do list—a list of tasks that you'll work on that day. Begin each study session by writing your to-do list. Treat it like a ritual. If you're not in the habit of writing a to-do list, here's how to get going.

1. The first item on your to-do list is always "Write today's to-do list." It's a task that must be completed.
2. Look at yesterday's to-do list, and add unfinished items to today's list.
3. Look at your calendar for assignments and add things to your to-do list as appropriate. If you've done a good job of maintaining your calendar, the entry for the current date should include reminders about upcoming work, e.g., "Political science test one week from today, chapters 7–11." This is a good time to double-check that you've flagged those upcoming assignments. Check the next couple of weeks to be sure.
4. Separate larger tasks into bite-size pieces as appropriate (more on this in chapter 11).
5. Scan your to-do list and decide on the order in which you want to do the tasks on it.
6. If, as you're working, you notice a new task that needs doing,

add it to the list. For example, if you're trying to write part of your study guide in preparation for a test and you see that you forgot to reorganize the notes from a lecture, add that to your to-do list.

This set of steps makes writing a to-do list look like a bigger deal than it really is. It shouldn't take more than ten minutes, and in the end **it will save you time** because you always know what you should do next. Without the to-do list, every time you complete a task, you must ask yourself, "Okay, now what?" Rather than having to make that decision multiple times, it's more efficient to decide once: *these* tasks, in *this* order.

This isn't a list of "things to do today." **It's a ranking, by importance, of tasks.** Typically you won't finish all the tasks on the list. And if you do, that doesn't mean the session is over. It just means you should look over your calendar to find out what to do next.

Using a to-do list will remove one possible source of study stress. As you work on one task, you will not wonder whether there is something more important that you've neglected. You'll know that you've evaluated all of your near-term tasks and that you're working on the most important one.

To-do lists also help fight a motivational problem. If you're like me, you sometimes finish a work session and feel as though you got nowhere. Maybe a series of small, unexpected problems arose, and you had to spend a lot of time solving them. Or a task that you thought was complete actually needed more work. In other words, you had setbacks that meant that even though you worked hard, you ended the session pretty much where you began. A to-do list won't magically put an end to that sort of thing, but you can at least look at your list afterward and say, "I'm not where I thought I'd be, but all that stuff had to be done." A to-do list encourages you by showing you what you've accomplished. For that reason, **review your to-do list at the end of every work session**. Make it a little ritual to take some well-deserved pride in all that you achieved.

The final way a to-do list can be helpful is by defeating your tendency to procrastinate. But for that purpose you need to write the list in a particular way, so I'll leave that discussion for chapter 11.

In a sentence: Creating a to-do list for each work session will help keep you focused, reassure you that you're working on what's most important, and show you what you've accomplished.

TIP 65

Set and Revisit Your Learning Goals

Most of this book concerns effective learning over the course of weeks, but part of planning is the intelligent selection of what to learn over the course of years. For some students, this choice sneaks up on them because they don't recognize their increasing responsibility for their own education.

Yet **without reflection and planning, you may miss important opportunities**. A high school student who doesn't like math may stop taking courses as soon as she's filled the minimum graduation requirement, only to discover, upon exploring college choices, that the schools of design that intrigue her expect a strong math background.

So **keep a list of your long-term goals**. What sort of work do you hope to be doing ten years from now? There's no need to be specific if you have no idea, but think in terms of broad categories: business, something mechanical, something artistic? Is the field you're interested in compatible with your ideal family life? Can you pursue the career you'd like anywhere, or would you need to go wherever the jobs are?

In addition to your goals, **jot down what you need to learn** to achieve them. The idea is to plan backward: I want to end up *there*, and to get *there* I need to do *this*, and to prepare for *this*, I need first to do *that*, and so on. Then **write down one or two specific steps** you might take to lead you closer to your goals: talking with an expert, perhaps, reading a relevant book, or taking an online course.

I'm not suggesting that you come up with a rigid life plan. I think Winston Churchill had it right when he said, "Plans are of little importance, but planning is essential." Churchill meant that the specific plans you draw up will almost certainly have to change because circumstances will change. But you will still benefit from planning because you will have thought about your goals, your capabilities, and the resources available to you.

That advice—plan but be flexible—applies not only to the goals you're striving toward but also the steps you need to take to get there. The college students who get into the worst academic jams are those who create a plan and stubbornly stick to it when it's obviously not working. For example, a student takes an excessive course load, so he ends up failing one class. His response is to think, "Oh no, I'm behind!" So the next semester he takes even more credits in an effort to catch up, thinking, "I'm going to stick it out. I'll just work harder!" You can predict what will happen.

In addition to flexibility, put a good dash of skepticism into your planning. The internet is wonderful, but you know it's not fully trustworthy. If you google "How can I be a software engineer?"—or a professional baseball player or a psychology professor—the website you will land on is unlikely to have been written by someone in that profession. It's written by someone trying to make a buck. **Supplement what you**

find online by talking with people who actually have the job you're aiming for. Students may feel awkward making this request of someone they don't know, but they shouldn't. People generally enjoy talking about themselves and want to help. That said, people are also busy, and some of us get more requests for this sort of thing than we can possibly fulfill. Be prepared for the first few people to say no before you get a yes, but rest assured that you're not making an inappropriate or strange request.

Revisit your list of goals every six months or so. Do they still hold? Six months ago, what steps did you say you would take next? How did they work out? Is it time to rethink your path or to gather more information about what you should do next? Research shows that **people who monitor their progress are more likely to achieve their goals** than those who do not. That's one reason it's important to write down the specific steps you plan to take; it makes it easy to assess whether you're doing anything about your goals.

The years pass more quickly than we anticipate. Make the most of them by investing a little time in long-term planning of your learning and career goals.

In a sentence: Set long-term learning goals related to your career aspirations, and revisit them every six months to see how you're progressing and whether they should be adjusted.

TIP 66

Set Goals with the Hidden Factors in Mind

How should you set long-term professional goals? You'd expect that three factors should be taken into account: (1) what you'll find satisfying in the long term, (2) your capabilities, and (3) the market. It seems obvious that all three matter. For example, you may think that a career as a musician would be quite fulfilling, and you may be quite capable (say, in the top 5 percent), but the job market for professional musicians is very small.

The stock advice is to "follow your passion"—that is, put the greatest weight on what you'll find satisfying. This advice often acknowledges the market factor by suggesting you channel your passion toward a job with a good number of openings. If you love music, for example, you might think about event planning or working as a music therapist.

Those offering the stock advice admit that it's still tricky to calculate the balance among these three factors, but that's not the only challenge. Other influences can warp your planning. Let's make them explicit and figure out how to address them.

First, "Follow your passion" is slightly off. **"Follow your purpose" would be better.** Research indicates that the happiest people are those who find purpose in their work, meaning that they perceive their work as having a positive effect on the lives of others. Passion might contribute to purpose, in that you're probably most likely to feel that your work has purpose when you're passionate about what you're doing—it's something that matters to you and that you feel ought to matter to others. Still, it's purpose that should be foremost in your mind.

Another drawback to focusing on your passion is that it encourages you to ignore your faults, and they can be informative. Activities that eat up your time tell you what you love to do, which may be the way to direct your search for purpose. If you think you talk too much, maybe you should be in a job that requires a lot of talking, like teaching. If you always have to be the center of attention, maybe you should be making sales presentations to large groups. If you see the negative possibilities in every situation, maybe you should be in risk assessment. People often want to eliminate their flaws, but sometimes our flaws are deeply embedded in who we are. It makes more sense to redirect them.

The second shortcoming of the usual formula for thinking about your goals is that it ignores your surroundings. Your goals are personal, but that doesn't mean other people won't affect the likelihood that you'll meet them. You are most likely to befriend the people you work with or attend school with. They are the ready means by which you'll learn new things and find new resources. In short, **your environment can be supportive, neutral, or toxic to your goals**, even if the people around you are unaware of your plans.

For example, when I taught at Williams College in the early 1990s, a very high percentage of the students jogged. Many I talked to said they hadn't been runners in high school, but when they had come to Williams, they had found that everyone jogged, so they did, too. Surely social pressure was part of the reason, but in addition, jogging at Williams was *easy*. It was easy to find someone to run with, to get advice about good routes and equipment, and to join a runners' club.

When you're selecting an environment—choosing a college, for example, or considering a job offer—you should be thoughtful about the environment you'll be joining. Do the people there seem to share your goals? Does the institution? Most important, is there real evidence that they do, or is it just something they say? For example, many companies say they support individual growth and encourage employees to learn new skills and grow into new jobs. Have there been examples of that learning and growth during the last few years in the division you'll be joining? Are there company policies supporting it, such as paid tuition for relevant classwork or paid days off for professional development?

Third, **people tend to underestimate how much their emotions can cloud their calculations**. Even if you think that passion should be more important than I have allowed, you still want your goal setting to be realistic. Your passion for stamp collecting, for example, shouldn't make you irrationally optimistic about the money you'll make if you start a business selling supplies to collectors.

There is a surprisingly simple and effective way of dealing with this problem. Each of us is much more clearheaded when offering advice to other people. Think about how frequently a friend describes a life choice that has him worried and confused but for which you think the answer is obvious: *No, you most certainly should not marry your fiancée who has cheated on you three times.* You can be objective because you're not emotional about the choice. You're not in love with her.

To be sure that emotions are not preventing you from thinking clearly about your own life goals, **try giving yourself advice as though you were someone else**. Talk about yourself in the third person and describe, aloud, the situation that you're in: "Well, Dan, you'd like to apply to transfer from the School of Engineering to the School of Education so that you can become a high school physics teacher. Let's start by listing what you've done to make yourself attractive to the admissions committee at the School of Education, and then we'll consider how selective the School of Education is and what your job prospects might be with that degree. And then we'll talk about the advantages and disadvantages of your current degree program."

Setting goals is complex, and it's subject to influences you may not recognize. The importance of the consequences means it's worth being as thoughtful as you can about the process.

In a sentence: When considering long-term career goals, be sure that they will contribute to a sense of purpose, recognize how your surroundings will affect your ability to achieve them, and be sure that emotions have not influenced the goals you set.

TIP 67

Develop a Plan

I've suggested that you make your goal specific and plan the first steps. But how can you maximize the chances that you will actually follow through and take those steps? These sorts of goals are difficult to meet because they feel like an add-on, something extra that would be great to do but are not part of your current responsibilities. Nothing bad will happen if you fail to do them.

Researchers have discovered a couple of ways you can make it more likely that you'll follow through.

To begin with, **make your plan even more specific**. Rather than saying, "I want to talk to antiques store owners in the next six months," you might specify a more detailed timeline for this work, starting with the idea that in the next month you will schedule one conversation and you'll plan to get the name of the next person to interview from the first. If I'm an aspiring real estate agent, I might plan that in the next month I will invite to lunch an acquaintance who is just getting started in real estate. The month after that, I'll identify three online introductory courses on the topic.

In addition, **have a plan B**. If I'm having trouble finding an antiques dealer who is willing to talk to me, I'll contact my aunt, who is active in the business community in her town, to see if she can help. If my real estate agent acquaintance isn't very helpful, I'll tell him that I'm trying to get as much information as I can and ask him whether any of his colleagues might spare me fifteen minutes to chat.

These contingency plans are for external obstacles, but it's even more important to **plan for internal obstacles**, things about yourself that may prevent you from following through. For example, suppose that you would have no difficulty asking your real estate agent acquaintance to lunch, but you never know how to steer a conversation to the topic that really interests you. Again, the solution is to formulate a plan in advance. Research indicates that **it's helpful to make your plan in an if-then form**. For example, you might think, "There will be a natural break in the conversation when the server comes to take our order. If I haven't brought up my plan by then, I'll do so just as the server is leaving."

Why does specificity help—both specificity of the plan and specificity of the contingency? It's another example of the strength of memory and the comparative weakness of our ability to solve problems. At the moment I decide, "This is my goal," I have lots of energy and mental space to think about steps to get there. Later, when I recall that I set this goal, I might be tired or less motivated, and in that state, I'm less able to solve problems. But memory is not much affected by mood or energy level. So if I made a plan earlier, I remember it and know what I'm supposed to do. The difference between problem solving and memory is even more important when obstacles pop up. If you're at lunch with your real estate agent acquaintance and you can't bring yourself to raise the topic that was actually the point of the lunch, you won't be able to devise a plan—you'll feel too nervous. But despite your nerves, you can recall a plan you made earlier.

In a sentence: You can increase the chances that you will pursue your goals by planning the specific next steps to take, anticipating obstacles (internal or external) that might prevent you from taking those steps, and creating an action plan to be taken if an obstacle arises.

For Instructors

Most teachers already help students with the job of remembering assigned tasks by encouraging them to use a planner. Fewer ask students, "How long do you think this assignment will take?" If you don't ask your students to plan that aspect of assignments, you might consider adding it and modeling your thought process for the estimation.

That still leaves your students the important work of pulling together assignments from different classes, making decisions about the relative importance of the competing demands on their time, and writing a to-do list. That requires coordination across classes, so a study hall or homeroom session may provide an opportunity for high school students to get some practice and instruction in this process.

You probably don't discuss long-term goal setting in your classroom, but the subject is fairly likely to come up in individual meetings with students. That's especially true when students face transitions: community college students wondering whether they should continue at a four-year institution, for example, or a college sophomore wondering whether a B in organic chemistry is a sign he should give up on medical school.

When you think a student's goal is well aligned with his record and prospects, these conversations are easy, even fun; you play the role of cheerleader and offer some counsel. But if the goal seems unrealistic, it's awkward. You want to provide honest feedback, but you also want to be supportive. So how sure should you be that a student can't meet a goal before you puncture the dream? And what should you say?

I've resolved this dilemma by not providing feedback in those terms. I emphasize what they've done relative to what's needed. I'd say, "Students who are admitted to medical school have usually done X, Y, and Z. You've sort of done X, and you've definitely done Y, but you haven't done Z." Then we discuss what it would take to do Z and, if she can't do it, possible ways around the requirement, if there are any. For example, if she's in her last semester of college, it's too late to remedy a low grade point average. But a couple of years as a research assistant in a scientific laboratory would provide good experience, put some distance between her and the low GPA, and earn her a helpful letter of recommendation from the head of the lab.

Summary for Instructors

- Help students develop the habit of keeping a calendar up to date.
- Most students would benefit from guidance about setting work priorities and budgeting their time.
- When discussing students' long-term goals, focus on what it would take for them to reach their goal and what they've done thus far, rather than making a global judgment about their talent or skill.

CHAPTER 11

How to Defeat Procrastination

Procrastination is challenging to avoid, but the psychology behind it is not complicated. **We procrastinate to make ourselves feel better;** we put off an unpleasant activity (for example, doing a math problem set) in favor of a pleasant activity (for example, playing a video game). Unsurprisingly, the more we dislike the task or the more appealing the alternative activity, the more likely we are to procrastinate.

But the problem is a little worse than it first appears because pleasure or pain that we're contemplating in the future doesn't have the same power as pleasure or pain *now*. For example, suppose your doctor tells you that you need to keep an eye on your sugar intake, and I ask you, "Would you like to have a piece of cheesecake for dessert a week from now?" It would be pretty easy to say, "No, I'm supposed to limit my sugar intake." The pleasure of the cheesecake doesn't tempt you much because that pleasure is a whole week away. But think how much harder it will be to turn down the cheesecake if I offer you a slice *now*. In the same way, painful things don't seem as scary when they are in the distant future. Even if going to the dentist frightens you, you might be talked into an appointment for a checkup if you're expecting to go six months from now. But what if the receptionist said, "Actually, there's been a cancellation . . . would you like to come right over?"

The way that outcomes change value over time helps us understand why playing the video game now (with the idea that you'll do your math problem set later) is much more appealing than doing your math first and playing the game afterward. Playing the video game now has a lot of positive value, but playing it later has less; it changes value over time, just like the cheesecake does. So does the math problem set. Doing it now seems really negative; doing it in the future, less so.

Impulse control also plays a role. An impulse is a plan your brain creates that meets an immediate desire but has bad consequences in the long run. When you see "Death by Chocolate (serves two)" on the dessert cart, you may have the impulse to order it. If someone cuts you off in traffic, you may have the impulse to run them off the road. People vary in how well they can control their impulses, and that's a big factor in procrastination.

To reduce procrastination, we can focus on (1) **making work seem more favorable compared to the alternative** and/or (2) **reducing the chances that we will act on impulse**.

WHEN YOU'RE TEMPTED TO PROCRASTINATE

What your brain will do: It will judge that the work you need to do will be unpleasant but less so later; further, an alternative to the work seems very attractive now but will be less attractive later. So work is put off, and the fun alternative is selected.

How to outsmart your brain: Make work seem less disagreeable, and make the tempting alternatives to work seem a bit less fun; it's all in how you talk to yourself about them.

As we'll see in this chapter, **your ultimate goal is to defeat procrastination by making work a habit.** If you sit down to your daily work session as automatically as you brush your teeth before you go to bed, you won't procrastinate—you've eliminated the possibility of choosing not to work because you're not making a choice. You're on autopilot. The hard part is consistently avoiding procrastination until you get to the point of work feeling like a habit. The tips in this chapter will help you get there.

TIP 68

Don't Rely on Willpower to Reduce Procrastination, Rely on Habit

When you get up in the morning, you don't think carefully about whether there is a more efficient way to make your coffee. You don't experiment by brushing your teeth with your nondominant hand. There is a lot of your day—probably most of it—during which you're on autopilot. You do things the way you always do them.

This isn't laziness. When you experiment with something new—"Hey, what if I replace my paper coffee filter with a lettuce leaf?"—the result is sometimes good but is often a failure. Doing a task on autopilot is not creative, but if the outcome in the past has been acceptable, being on autopilot means you'll get the acceptable outcome again.

More important, for truly habitual actions we not only have an unthinking, routine way of doing them, we often don't need to think about starting that routine. You don't walk into the kitchen in the morning and think, "Hmm. Should I make coffee?" You just make it. You don't procrastinate doing routine actions because *there's no act of choosing.*

I've suggested that you pick a set period of time during which you will work each day (see tip 62). Ideally, starting your work session will become as habitual as flossing before bed. **When sitting down to a work session becomes habitual, there's no chance of procrastinating, because you're not making a choice.**

How can you make an action into a habit? As you've probably guessed, consistent repetition is the answer, but if you ensure that the repetition has a few key features, the habit will develop more quickly.

First, **it's easier to establish a habit as a sequence of things you do rather than at particular times**. Habits are like memories in that they are cued. Something happens in the environment (or in your mind) that cues a mental action plan: "Do this now." You have a routine in the shower—a sequence in which you wash your body, shampoo and condition your hair, shave, whatever. Finishing one action in your shower routine cues the next one. The cue is *not* "It's 6:35 a.m." **Time is a bad cue** because you don't monitor time that closely. In contrast, completing an action is obvious to you—it's hard to miss that you've just rinsed your hair.

To develop the habit of working, start by considering what might serve as a cue. If you're a high school student, maybe it's "I've finished cleaning the kitchen after supper" or "I've finished my after-school snack." You must ensure that the trigger is certain to be done each day. Thus, "I've finished the supper dishes" isn't a good trigger if you alternate cleanup days with your sibling.

Another way to speed the development of a habit is to **choose the context wisely**. Schedule your consistent time for studying in a part of your day when you *can* be consistent. Don't schedule it for "when I get home from school" if you frequently want to socialize after school. But note that it's fine to set your time as "after my Saturday workout" even if your workout time varies. Just as you can wake up anytime and stumble into the shower on autopilot, the routine will be the same as long as the cue is consistent—arriving home from my Saturday workout.

How long does it take to develop a habit? In one experiment researchers paid subjects to develop a habit of their own choosing that related to healthy eating or drinking or to exercising. The new behavior felt habitual after an average of 66 days, but that figure varied a lot—between 18 and 254 days. The number surely depends on the particular habit you're trying to develop, your personality, and the fit between the two.

If you're a college student, I suggest you try thinking of college as a nine-to-five job (or ten-to-six, or whatever works for you). Treat those forty hours from Monday through Friday as nonnegotiable work time. You don't do laundry or socialize during that time. You're in class or studying.

When my fellow professors and I are thinking about who to admit to our PhD program, we have a slight bias toward people who have been in the working world compared to people coming straight from college. That's because holding a nine-to-five job makes working habitual; you get used to showing up and working, even if you don't much feel like it on a given day.

Treating your learning time as a habit sounds great, but you still have the on-ramp of sixty-six days (or whatever it ends up being) when you *do* have to rely on willpower. Willpower is an unreliable ally in ensuring that you work. It fluctuates with your mood, your physical state, and the environment. Let's look at ways to make sure you work consistently during your set time so that the habit has a chance to develop.

In a sentence: Making your work session habitual is the ultimate way to defeat procrastination because it removes the need to choose to work.

TIP 69

Each To-Do List Item Should Be Concrete and Take Twenty to Sixty Minutes

When eating an elephant, take one bite at a time. —Unknown

A journey of a thousand miles begins with a single step. —Chinese proverb

One day at a time. —Alcoholics Anonymous slogan

Each of these quotations makes the same point: ambitious goals are so intimidating that we won't attempt them. The trick is to set a much smaller goal. Don't think about eating the whole elephant, take one bite. Don't think about refraining from alcohol the rest of your life, just make it through today without a drink.

Here's why this strategy works: when you make a choice, you consider not only how much you like (or hate) what you'll get but also the odds that you'll get it. For example, if you offer me the choice of a chocolate bar or $100,000, it's pretty obvious which one I'll take. But suppose you say that if I choose the chocolate bar, you'll definitely give it to me, but if I choose the $100,000, I'll have a 0.000036 percent chance of getting it. In more everyday terms, you give me a dollar with which I can buy a chocolate bar or a lottery ticket. I love the idea of getting all that money, but if I make that choice, I'm very unlikely to get it. I'd rather have the sure-thing chocolate bar.

We are more likely to procrastinate if we think we can't succeed at the task we ought to do. If your instructor assigns *Bleak House*, you not only have all of the usual reasons to procrastinate; you also notice that the book is more than nine hundred pages long. Feeling as though you can never finish such a long book makes starting it feel like buying a lottery ticket. "The prize—finishing the book—sounds appealing, but I don't think I'll get the prize. So why start the book?"

The quotations all offer the same advice: break overwhelming tasks into small, achievable bits. The title of Anne Lamott's classic book on writing, *Bird by Bird*, came from an overwhelming homework assignment. She explained:

> Thirty years ago my older brother, who was ten years old at the time, was trying to get a report on birds written that he'd had three months to write, which was due the next day. . . . He was at the kitchen table close to tears, surrounded by binder paper and pencils and unopened books on birds, immobilized by the hugeness of the task ahead. Then my father sat down beside him, put his arm around my brother's shoulder, and said, "Bird by bird, buddy. Just take it bird by bird."

Tip 64 suggested that you write a to-do list at the start of each study session. **Each item on your to-do list should be a small bite—shoot for twenty to sixty minutes.** Many learning tasks do not come in small bites, obviously. You need to disassemble them into parts, but you may not know how to. **If you don't know what the pieces should be, make that an item on your to-do list.** It's work, and it might take a while, so write, "Figure out plan for economics class project."

Let me offer a little assistance on breaking big tasks into smaller pieces. I can suggest three possible principles.

Some tasks are best thought of in **phases** or steps, with each phase depending on the outcome of the previous phase. For example, a project report has four distinct phases: research, outlining, writing, editing. In chapter 6 I suggested these steps in preparing for an exam: create a study guide, commit your study guide to memory, meet with your study group, and overlearn—that is, keep studying even after you know the content. (Each of these phases is actually a sizable task that you'd want to break down further.)

Other tasks are ordered not in sequenced steps but rather in **categories**. That's the breakdown Anne Lamott's dad suggested with his bird-by-bird advice. The same principle applies within the "create a study guide" phase of studying: you would write the part of the study guide that covers lecture 1, then lecture 2, then lecture 3, but you could also write them out of order if you preferred.

Other tasks break down naturally into **parts**; the task is really one giant thing to do, but you create artificial pieces to make it more manageable. When you're at the writing phase of a project report, you can break that phase down into parts. You can consider an assignment with fifteen problems as composed of three five-problem chunks.

Whether you divide your task into phases, categories, or parts, be sure that your description is as concrete as you can make it. The goal is that when you undertake a task, you don't need to think through what you're supposed to do. Don't write the task "Review for Govt. quiz." Review how? Reread chapters, read notes, make an outline, what?

Of course, you want the task to be not only concrete but also relatively short. I've said twenty to sixty minutes, but there's nothing sacred or research based about that figure. Just keep the goal in mind: you're trying to trick yourself into working by making the task seem easy, harmless. A small bite.

In a sentence: Make each to-do list item doable—between twenty and sixty minutes long—because procrastination will be less tempting if tasks look achievable.

TIP 70

Reframe Your Choice

Redescribing your choice may also make work seem more appealing. To see how this strategy works, we'll use an idea economists call *opportunity cost*. It basically means giving up the chance of a potential gain.

For example, suppose your fabulously wealthy aunt gives you $100,000 when you're seventeen, no strings attached. You could keep the $100,000 and get a job immediately after high school graduation. Or you could spend the $100,000 on college, figuring that the expenditure is a good investment because you'll probably land a higher-paying job if you have a college degree.

Tuition is obviously a direct cost, but going to college also includes opportunity costs. You lose the chance to increase your $100,000 by investing it. In addition, attending college means you lose the possibility of having worked for four years, and during that time you would have earned an income, possibly some retirement benefits, and a reputation as a good worker, someone deserving of promotion.

Now, what does all this have to do with procrastination?

Suppose you're a college student, it's Thursday night, and your roommate asks if you want to watch a movie. You were thinking you'd work on a chemistry problem set that's due Monday. The natural way to think about this situation is similar to the video game/math example problem provided earlier: immediate fun and boredom later versus immediate boredom and fun later. But choosing the movie carries an opportunity cost you might not have considered: if you go to the movie, you will not have the satisfaction of finishing the problem set.

The next time you are tempted to procrastinate, **try describing the**

choice to yourself in a way that highlights the opportunity cost. Don't ask, "Shall I watch a movie or shall I work on the problem set?" Say, "Shall I knock that problem set off and be done with it, or shall I put it off and give up my chance to feel good about having it over with?"

Here's another way to reframe the dilemma that you may like. The psychologist Alexandra Freund has pointed out that we tend to dwell on one of two things when we procrastinate: either we don't like the process the task involves, or we don't like the goal. For example, a student may procrastinate about writing a study guide: it's not the creation of the study guide that she minds (the process), it's taking the test (the goal) that she hates, because she has test anxiety. So she puts off doing anything associated with taking the test. Another student may hate doing data analysis but enjoy designing and delivering a PowerPoint presentation for the class once the data have been analyzed.

If you notice that either the process or the goal is the part of the task that makes you procrastinate, see if you can **focus on the part of the task you don't mind doing so much**. The first student might try telling herself, "I'm not taking the exam. My job now is to summarize what I've learned." The second student might say to herself, "This isn't just number crunching; what I'm doing is preparing for my presentation." See if reframing the task to highlight the part you enjoy (or at least don't mind so much) makes you less likely to procrastinate.

In a sentence: Redescribing the work ahead—the outcome of the work, the process, or the goal—might make the right choice more appealing.

TIP 71

Just Start, and You'll See That It's Not That Bad

People are surprisingly poor at predicting their emotional reactions. Sure, you know that if you overhear someone say that you're cute, it will feel good, and if they say you have a terrible sense of humor, you'll feel hurt. People usually get the direction of their reactions right (positive or negative), but they overestimate their strength and duration.

Psychologists have examined this overestimation of emotions as it relates to exercise, an activity that often makes people procrastinate. They found that one of the reasons people put off exercising is that they think they will feel more miserable when they exercise than they actually do.

You may find that the same is true of mental tasks. If you can just get yourself started, you'll see that **working really isn't as unpleasant as you thought it was going to be.** One way to convince yourself to "just start" is to tell yourself, "I'll just work five minutes. If I really hate it, I'm allowed to stop."

When my sister wanted to develop the habit of jogging every day, she came up with a similar strategy. If she put on her running clothes and jogged to the end of her driveway, that "counted." She was allowed to quit but still tell herself, "I jogged today." Of course, 95 percent of the time, even when she had thought, "I do *not* feel like jogging today. This is a day I really will turn around at the end of the driveway," she kept going when she got there. It just didn't feel that bad.

If you're having a terrible time getting yourself to sit down and work, try telling yourself, "I'll make my to-do list for today. **If, after I make my**

list, I want to take a break, I'll take a break." Once your to-do list is written out, there will probably be one or two things on it that don't seem too hard to tackle. And you're off.

But. For this tactic to work, it's crucial that you really do allow yourself to take the break if you want to, once the to-do list is written. The whole point is to make starting seem unintimidating by giving yourself permission to ease off quickly. If you know that the permission is a lie—for example, that you'll feel guilty if you take the break—you're not making it unintimidating.

In a sentence: Starting a work session will seem less odious if you give yourself permission to take a break after a short time.

TIP 72

Tell Others What You're Up To

Humans are intensely social beings. Much of what we do, we do with others, and even when we do something on our own, we consider how others will view us: Will we make the people who matter to us proud or angry or happy?

If you are frustrated by the frequency with which you procrastinate, you can take advantage of your social connections to help you work on the problem. It's a start simply to tell friends, "Hey, I'm trying to procrastinate less and do a better job of keeping up with my work." The hope is that your friends, knowing of your plan, will help in two ways: they will make you accountable, and they will provide support.

Shame is a key reason that accountability works. You feel embarrassed if you've told people, "I'm going to stop procrastinating!" and then, within a week, you've obviously not stuck to your resolution. There's no shame in using shame as a prod, and if you like that idea, you can put even more on the line by using a commitment website such as stickK, 21habit, or Beeminder. Sites come and go, of course; search for "commitment contract websites" for the latest.

Most sites have similar formats: you commit to doing something—say, "Work from 7:00 to 10:00 p.m. each weeknight." Each day you log in to the website and report whether or not you met your commitment. (Some sites require that you have a referee to monitor your honesty.) If you don't meet your commitment, a certain sum from your credit card goes to a nonprofit organization. You're encouraged to pick a cause that makes the loss more unpleasant, for example, one with a political view you dislike. For added motivation, many websites broadcast your failure via your social media account.

Participating in a study group (see tip 23) also has built-in accountability. When the group meets to compare lecture notes or discuss an upcoming test, the other group members are counting on you to have done the preparatory work. The responsibility to your peers may help you meet these deadlines.

In addition to holding you accountable, **friends can provide positive support**. Depending on the circumstances, we might need one form of support more than another. Psychologists list four types:

Emotional support: People who express sympathy and caring. For example, when you feel frustrated because you still procrastinate even though you try not to, your friends can listen to you sympathetically and encourage you to keep trying.
Informational support: People who offer advice, suggestions, or information. For example, people might offer you their own strat-

egies for defeating procrastination or help you find a commitment website you like.

Practical support: People who do things that directly help your effort. For example, a friend sticks up for you when someone gives you a hard time for saying you are going to work instead of going to a party. Or a friend offers to check in with you to be sure you're studying the afternoon before a test.

Appraisal support: People who provide information to help you self-evaluate. For example, a friend provides an objective view of whether your efforts to procrastinate less are working. A friend might also remind you of other times your determination to change something about yourself paid off, giving you more confidence that you can defeat procrastination.

When you read about these types of support, I expect a couple made you think, "Nah, that's not what I need my friends to do," and one or two made you think, "Yes!" That realization might make you selective about who you seek out for support. Obviously it's a bad idea to talk about your antiprocrastination campaign with the worst procrastinators you know. But after those folks are ruled out, look at these different types of support, and **think a bit about which of your friends are most able and willing to provide the type of support you need**.

But don't expect your friends to automatically know how to help; you probably need to tell them, and the list above can help you be more articulate in describing what sort of support you're hoping to get and what sort of concrete actions they can take.

Some people find it hard to ask for help. Bear in mind that you probably don't think less of people when they ask for your help. Most people are happy to assist their friends; it makes them feel good, so don't deny others that opportunity.

In a sentence: Your social network may provide emotional support and practical help in your effort to procrastinate less, but to do so, people must know that you're working on it and what sort of help you need.

TIP 73

Consider Whether Your Procrastination Is a Way to Self-Handicap

In my sixth-grade shop class we made birdhouses. After the first work session, I concluded that I was Bad at This. I kept getting confused about how to fit the template onto the plywood, and my saw cuts weren't straight. I soon made a show of not trying. I worked quickly, I didn't pull and redrive nails that bent, and so on. I was making a preemptive excuse for poor work.

Psychologists call this *self-handicapping*: we impose a handicap on ourselves so that we have an excuse for our failure.

Procrastination makes it easy to self-handicap but deny what you're up to. You don't say, "I'm not going to study for this exam, because if I study and fail I'll feel stupid." Instead you fill your time with other things (laundry, socializing, other work), and somehow you don't get around to studying until the last minute. And there's your excuse: "Oh no, I was so busy I barely had time to study for this!"

Why would you do that? Doesn't it make sense to try your hardest? Even if you think the chances of earning a good grade are low, at least you can improve the odds by studying. Self-handicapping must mean that getting a low grade after you studied reveals something deeply troubling. What could it be?

The answer, of course, is "stupidity." In tip 60 I mentioned that many people believe that intelligence is largely genetic and largely unchangeable. They further believe that, being genetic, intelligence is a matter of what you are, not what you do; that is, smart people didn't get that way by working hard but by having good genes. So if you need to work hard to pass a test, that shows that you're not very smart. Imagine, then, what it shows about a person if he works hard and fails a test!

These beliefs about intelligence are false. Yes, intelligence is determined partly by your genes, but it also depends on what you do; intelligence can be improved, and learning is the way to improve it. Therefore, "If you're smart, you don't need to study" is wrong. So is "If you fail a test, that shows you're dumb."

So what's a better way to think about failing a test? There's no reason to think of it differently from any other challenging task. If you showed up for open-mic night at a comedy club and didn't prepare, you wouldn't get many laughs. Now, suppose you *do* prepare and you still don't get many laughs. Does that mean you are simply unfunny? Or does it mean that comedy is challenging and that you need to prepare carefully and expect that becoming successful will be a long process?

Teaching yourself to learn is not easy. Over the course of reading this book, you've seen that there are a lot of components you have to get right. Be patient. If you stick with it, you will see results.

In a sentence: Some people procrastinate as a form of self-handicapping, so that they have an excuse if a test or project goes poorly.

TIP 74

Make the Temptation a Reward

This strategy should be your last resort: compromise between the work you must do and an alternative activity that tempts you. Do some of each by making the tempting activity a reward for working.

This strategy might be especially effective when you judge that the thing that tempts you to procrastinate is time sensitive. To put it more colloquially, FOMO is involved. FOMO stands for *fear of missing out*, but here I'll expand the term to mean a more general *feeling* of missing out.

Maybe a new "Halloween update" for your favorite video game went live a few hours ago or your favorite team is playing and the game is televised. Whatever the tempting alternative, the problem is not that you dread the work you're supposed to do or even that the tempting alternative is *that* amazing; it's that if you work, you're going to feel that you're missing something you cannot possibly get later. What should you do?

First, let's note what you should have done. A time-sensitive, attractive activity is usually something planned, something you know about in advance. As noted in tip 63, you should put social events into your calendar, not just work stuff. When you hear on October 16 that the Halloween update is coming on October 30, block out a couple of hours at the relevant time. **Protecting time for important social events will reduce your procrastination.**

Okay, you didn't protect this time, and now you've got a couple of hours of work you really need to do. If the tempting alternative can be packaged into time-limited chunks, you can **do a bit of the fun activity as a periodic reward for working**—say, work thirty minutes, then treat yourself to five minutes of the game. Note that I'm drawing a distinction between wanting to do something because it's special *now* and generally having a problem focusing your attention once you've begun to work because you'd rather be playing a game or checking social media. We'll address that problem in the next chapter.

This strategy is a last resort, because you run the risk of a "break" extending much longer than you'd intended. If you're teetering on the brink of ignoring work altogether in favor of the tempting alternative, deploy this last-ditch tool.

> *In a sentence:* If an activity is so tempting that it will make you skip your work session altogether, make that activity a reward for work completed.

TIP 75

Track Your Progress but Ignore Your Streaks

The key to developing a habit is consistently working at the time planned every day. If you keep track of how frequently you stick to this plan, you can take deserved pride in your dedication. It's easy to keep track if you **make a tick mark on your calendar for each day worked**.

There will be days when you sit down at your work space and feel unmotivated and pessimistic about getting anything done. Sometimes you'll be wrong, but other times the feeling will be accurate. You can't concentrate, and you have the terrible day you anticipated.

There are two reasons why showing up every day matters. First, even when you foresee an unproductive day and you're right, **something is much more than nothing**. You're still making progress. Second, and more important, you are showing yourself that the work matters to you. I discuss self-image more in chapter 13, but for now, consider this: When you see someone who sticks to her work schedule every day, even when she's tired or getting over a cold or just doesn't feel like working, what do you think? Obviously, you conclude, "That work is really important to her."

You draw the same conclusion when you see yourself work consistently. **Noting that you have met your work commitment builds your self-image as a learner.**

But there's an important distinction between taking pride in working consistently versus becoming obsessed with a work streak. Do not think to yourself, "Wow, I've worked 50 days in a row. I wonder if I can make it to 100. Or 365!" Monitoring streaks has a significant drawback: a single failure takes on undue significance. For example, a dieter succumbs to a fancy dessert at a wedding and so concludes, "I blew it. After all that work, I ate that rich dessert." So that night, when she's at home, she figures, "As long as my diet is shot, I may as well eat this ice cream."

In fact, not only should you avoid thinking in terms of streaks, you *should* take days off for things that are really important. If you're at a restaurant celebrating your girlfriend's birthday, you shouldn't be glancing at your watch and badgering her to finish so you won't be late for your study session. That's not showing yourself that you are dedicated to learning; that's showing yourself that your streak matters more than your girlfriend. **Be consistent, but don't prize streaks.**

> *In a sentence:* Taking note of the consistency of your work habits will be motivating and help to maintain them, but don't be fixated on streaks, because streaks are inevitably broken (and should be!); a broken streak will needlessly discourage you.

For Instructors

Procrastination is a nearly universal problem, and if you struggle with it, too, it may help to tell your students so. Some students see themselves as weak, saddled with a self-control deficiency that few others can understand. You can help students see that procrastination is a nuisance everyone faces and that successful people learn to manage. Students can learn to manage it, too.

Some of the tips listed in this chapter are good candidates for a bit of in-class modeling and support. For example, students are likely familiar with the idea of a to-do list, but they may not be experienced in creating one, and they may especially need help in thinking about how to break complex tasks into more manageable ones.

In addition, students will need help thinking about scheduling these subtasks. Instructors can help by providing concrete reminders. In other words, don't simply remind the class, "You're all working on your papers, right? They are due in two weeks." Instead, let them know what they should have completed by this point: "Don't forget, your papers are due in two weeks. You should have picked your topic, and you should have five of your ten sources identified. If you're not there, don't panic, but I want to see you right after class." In fact, you might consider not just providing such guidelines about desirable progress but actually setting interim deadlines.

Even for shorter tasks, students may be frozen by indecision because they are unsure of exactly how to proceed. You can help by making assignments as clear as possible. And think this through: If a student doesn't understand what's required or understands but has no idea how to get started, what are her options? Is it clear where she can get help? Make sure every student knows.

Summary for Instructors

- If you're troubled by procrastination, humanize the problem for your students by telling them so, and share how you deal with it.
- Help students think through how to prioritize tasks and break big tasks into smaller chunks.
- Set interim deadlines for large projects.
- Be sure that assignments are clear and students know what to do if they don't know how to get started.

CHAPTER 12

How to Stay Focused

Chapter 11 described procrastination and how to overcome it, but getting going is not the only problem; you also need to stay on task.

Let's consider a typical case of distraction. A student is studying, and her phone pings. It's an unimportant text message, but it's from someone with whom she has a Snapstreak that she hasn't maintained today. The next thing she knows, she's on Snapchat, extending that streak and others, then adding to her story, and suddenly thirty minutes have elapsed.

The psychologists Angela Duckworth and James Gross have described **four mental steps that apply to distraction**: first, the student *arranges her situation* for studying and it includes her phone; second, the student *shifts her attention* from studying to her phone when it pings; third, she *evaluates* the notification as important; and fourth, she *responds* by going on Snapchat.

The result—that she's no longer studying—could have been interrupted at any of the four stages. The phone could have been elsewhere and she would not have heard it. Or if the phone had been present, she could have ignored the notification. Or if she had paid attention to the notification, she could have decided it was unimportant. Or if she had decided it was important, she could have said to herself, "Even though it's important, I still ought to work now."

Note that **each change sounds harder than the previous one**. It's relatively easy to turn your phone to silent for an hour but much harder to keep yourself off Snapchat once you decide that your unfinished Snapstreak is important. Later in this chapter we'll consider a number of methods to interrupt distractions, focusing especially on those that are easiest.

I've described the loss of focus caused by distraction from the environment. **You can also lose focus due to mind wandering**, a term scientists use just as you do; it's when your attention, seemingly on its own, moves from your desired focus to something else. As you would predict, mind wandering increases (1) the longer you've been doing the task, (2) the more boring you find the task, and (3) if you consider the task really simple or really difficult.

Why the mind wanders is not well understood. Some researchers believe it happens when people judge current thoughts to be unimportant. Others believe that mind wandering is actually the natural state of the brain, which is why it takes effort to focus on one thing. Though we don't understand what causes it, researchers have identified some effective techniques to reduce it.

WHEN TRYING TO FOCUS YOUR ATTENTION

What your brain will do: It will direct your attention away from work when new information appears in the environment (distraction) or it will spontaneously redirect your attention to thoughts other than work (mind wandering).

How to outsmart your brain: To limit distraction, the easiest fix is to change your surroundings. Defeating mind wandering is harder, and the best strategy may be to accept its inevitability and to return promptly to the work at hand.

Different circumstances call for different techniques to stay focused, so we'll review several strategies, and I'll describe when to use each technique.

TIP 76

Choose Your Work Location with Care

"Find a quiet place to work." It's one of the most commonly offered pieces of advice for students, and with good reason. A location can facilitate concentration or bristle with distractions, yet once you pick a spot to work, you'll probably stay there even if you find it hard to concentrate; it's a pain to move. Not everyone has choices about where to work, but if you do, let's consider the features you should look for.

On the one hand, the characteristic you're after in a study space is pretty obvious: **find the place that is closest to being distraction free**. On the other hand, different environments offer different types of distractors, and these bother people more or less. A lot of my colleagues love to work at home because it's quiet and people don't drop in, but I get distracted by the kitchen—I wander in, open the fridge, and stare—and I'm distracted by all the little home maintenance and repair jobs I know I ought to do. A friend of mine told me he can't work in the same room as a bed or comfy chair because he'll talk himself into a "five-minute power nap" that turns into an hour.

You have to figure out what works best for you. When I was in college, I sometimes studied in empty classrooms, and I picked a different one each night (see tip 49). That was great, because a room containing nothing but a whiteboard and desks is *really* distraction free. But later I realized that I didn't like studying alone. **You may benefit from being around other people who are also working.** We are a social species, and we tend to feel and do what others around us are feeling and doing. It's called *social contagion*: if everyone around me is laughing, fearful, exercising, or studying, I will likely share those feelings or do those activities. Studying in the library meant being surrounded by other people working hard, and that inspired me.

In addition to *where* you work, **give thought to *when* you work**. Some people work best while others sleep, because a lot of distractions retire when people do. I do this myself, usually getting up around four. That schedule is not practical for most people, I realize—you need to have common sense about your body's sleep preferences, and, whatever your preferences, your schedule may have limits you can't adjust. But especially if you're a college student with a flexible schedule, think about going to sleep earlier and waking earlier on weekdays. It may make it much easier to find distraction-free time to study.

Finally, **don't be unrealistically optimistic about situations in which you think you can work effectively**. You may think you'll get work done while you're babysitting an infant ("He'll probably sleep a lot") or at your friend's intramural basketball game ("She mostly sits on the bench anyway") or in an airport or riding shotgun in your friend's car as you drive home together for Thanksgiving break. There's a temptation to think, "Oh, I'm going to be in this boring situation, I should do something useful." But sometimes you're bored in locations that are poorly suited to working. So, sure, bring a book for the layover; maybe you'll get some reading done. Just don't count on it.

In a sentence: The injunction "Pick a quiet place to work" is mostly right, but it's too simple; you should think about the best time, too, and about the possibility that other people may energize you, not distract you.

TIP 77

Improve Your Work Location

Sometimes you can't choose your location, or maybe you've chosen the best one available but it's still not ideal. Then what? You can make small changes that will reduce the chances that you'll be distracted.

Let's start with distractions in a classroom or lecture hall. You should try to **sit in the front row or near it**. That way, there is less chance that someone in front of you will do something distracting, both because there are fewer rows of people in front of you and because people who sit near the front tend to be more serious about paying attention. Furthermore, sitting close to the speaker allows you to see her facial expressions better, and that will provide a small assist to maintaining your attention.

What if you arrive too late to get a seat in front and someone in front of you with a laptop starts watching *New Girl* or shopping for vintage watches? It's going to be hard to ignore the activity on the screen, so you should **change seats if you can**. If that's not possible, at least try to shift your body in your seat to make the screen harder to see.

Suppose a friend keeps talking to you during the lecture. It's awkward to move or to tell them to be quiet. **Tell them that you're having a hard time catching everything** and see if that helps. If that doesn't work, say that you're going to start sitting closer to the front to be sure you can hear. That way you're not shutting them out, but you're making it awkward for them to keep chatting with you.

You face different distractions when you're working in a public place like a library or coffee shop. If the problem is just noise, **try wearing foam earplugs**. Another problem is that because you're studying in a public place, people might assume you're open to chatting. Try selecting a seat so that your back will face most foot traffic. Not looking up often will help, and wearing a cap pulled low makes it look to others like you are in your own world and will make them reluctant to invade it. If you **add noise-canceling headphones**, that should seal the deal.

But the greatest source of distraction may be electronic, not human. When I work on my computer, if I can see open tabs, documents, or folders, I will think of other work to be done or fun websites to visit. If you **use full-screen mode,** you can see only what you're trying to focus on.

Your phone is easier to deal with, because you can set it to silent. It's tempting to think, "I'll just ignore notifications," but the very act of ignoring one is an interruption. Thus, if you **turn your phone off**, it's better than setting it to silent; you will be less tempted to take a quick peek to see if you've gotten any messages. Schedule periodic check-ins during work sessions.

If you think you can be (or need to be) a little more intense about limiting your access to distracting digital content, you can **install a screen-time limitation app** that does it for you. Tools such as Freedom, AntiSocial, Cold Turkey, and SelfControl let you make a decision *now* about how much time you will allow yourself to spend using specific apps *later*.

If that approach seems too extreme, here are two other ways to limit your off-task time. First, you can **install a screen-monitoring app** that measures how much time you spend using different apps over the course of a week. Second, you might **turn off auto-login** for social media apps. The act of typing in your user name and password is a pain and the requirement will probably make you check in less often. You might also **turn off alerts** for some apps.

You can see two general principles running through these ideas. First, if something distracts you, remove it from your environment or at least make a change so it will bother you less. Second, if the distracting object is something you tend to seek out, make it harder to access.

The next tip further explores this second idea—the problem of distractions that you yourself seek out.

In a sentence: If a situation has something distracting in it, you can remove the distractor or make it less noticeable; if it's a distractor *you* seek out, you can make it harder to access.

TIP 78

Don't Choose Distraction

A friend of mine who grew up in Belgium and France described a difference in the work habits of Americans and Europeans: "Here, you tell yourselves you work all the time, but you're always doing something else while you work. In France, if you want to have an espresso, you have an espresso. Here, you get the espresso, and you work while you drink it. Or you pretend to listen to music while you're working. Or you put your feet up on your desk and say, 'Look at me. I'm relaxing, but I'm also working.'"

People are terrible at doing two things at once. Much—perhaps most—of the distraction people suffer is self-imposed. They just don't realize what they're doing to themselves. **Anytime you multitask, you are distracting yourself.**

It's obvious that you can't do a task—say, write a paper—as well if you're simultaneously doing something else, such as working math problems. It's less intuitive (but true) that there's also a cost when you switch between tasks, as when a student works on an essay while participating in one or more texting conversations. The cost is traceable to the difference in the "mental rules of the game" of the two tasks. For example, if you're texting a friend while writing an essay on *Invisible Man* for English class, you use one way of writing for the essay and a different way of writing when you text. When you switch between them, you need to do a mental reset and say, "Okay, now I'm writing *this* way." A great deal of research going back to the 1990s shows that this reset happens even if the tasks are very familiar and very simple, such as driving and holding a conversation. You'd think you could keep two very simple tasks in mind at once, but you can't. There's always a mental cost to switching.

Now, it's okay to say to yourself, "I don't care if I write the essay a little slower—I want to text my friend while I write it." Fair enough. I just want you to know that the cost is probably higher than you think. Laboratory studies show that even when people realize that there's a cost to multitasking, they consistently underestimate it.

That's especially true for media multitasking—that is, listening to music or having a video playing in the background while you work. Surveys of students show that they think this sort of multitasking doesn't get in the way of studying, and some say it even helps. But research tells a different story. Most studies have students come to a laboratory and bring the sort of music or video they play while studying. Then they are given a chapter of a textbook to read or some math problems to solve; some people have the music or video going while they work, and others don't. The results for video content are really clear: **if a video plays, work suffers** in time, accuracy, or both. That's true even when people feel as though they are ignoring it and it's just background noise.

Music, however, is more complicated. Researchers have tried all the variations you would probably think of: music with lyrics versus instrumental, classical versus pop, and so on. None seems to make much difference. Music sometimes helps academic tasks and sometimes hurts, because it has two conflicting effects: it's distracting, but it also has the potential to energize listeners; that's why people listen to music while they work out. Whether the combined effect is positive or negative depends on how energetic you're feeling, the difficulty of the task, and other factors.

When we add another task while we're working, it tends to be one that provides an emotional lift. We listen to music, text a friend, or check social media. Since these tend to affect your performance negatively, rather than seeking a boost by multitasking, **get your emotional lift during rest breaks**.

The bottom line on media multitasking is this: playing a video, even in the background, will interfere with your work. Music may be okay at times, but be cautious in drawing that conclusion. As I've emphasized throughout this book, people aren't great at evaluating their own thought processes or the quality of their work.

In a sentence: Don't multitask; you can't truly share attention between tasks, so adding a second task always compromises the first.

TIP 79

Rethink Your Evaluation

Recall the steps of mental control that apply to distraction from the start of the chapter: arranging the situation, shifting attention, evaluating, and responding. We've looked at ways to arrange a better situation, and we've thought about ways to make it more likely that you won't pay attention to a potential distractor. What about the evaluation?

Reevaluating a distractor might help with social media. When your phone pings and you reach for it, it means you have evaluated the ping as more important than whatever you're working on. But the truth is that you probably didn't really appraise it; reaching for the phone was automatic. One strategy you could try is to **interrupt that automatic act and consider it**. Think or, better yet, say aloud, "What are the odds that this notification is *really* important?" Indeed, how likely is it, if you wait until you take a break to view the message, that you'll regret it, that you'll think, "Why did I keep working? I should have listened to my instincts and checked my phone immediately." If reaching for your phone is so automatic that you feel you have no chance to pose such questions to yourself, you can put a safeguard into place by wiping your automatic login so that you must enter your login ID and password each time.

You can try a similar reevaluation technique for mind wandering. Say aloud, "I don't need to think right now about what I'm going to wear to that wedding next month. I will think about that as I'm driving home."

This technique is harder to pull off when the mind wandering content is emotional. For example, maybe you're upset because your manager cut your hours at work. You've got an exam coming up and you need to study *now*, but you're obsessing about the money problem. Sure, if you ask yourself, "Is it helpful to think about money now?" you'll say, "No," but it won't keep those distracting thoughts out of your mind.

You can try a technique I've suggested in a couple of other circumstances in which emotions cause a problem: psychological distancing. Think to yourself—or again, better, say aloud—what you think you ought to do, **but talk about yourself in the third person** (ideally in a place where other people can't overhear you). For example, say, "Dan is really upset right now. He's very worried about money, because he's lost some hours at work and he's not sure how he'll make his rent. But there's no point in him thinking about that now. Dan has time tomorrow afternoon to go to the student employment office to look for another job. And he can take another look at his budget then. But he can't do those things right now. Right now, Dan needs to prep for this exam."

The idea is that talking about yourself in the third person gives you some emotional distance from the problem and makes it easier to respond to a difficult situation in what you rationally know is a useful way.

The techniques described here focus on reevaluating a potential course of action: Do I really want to check my phone now? Do I really want to obsess about my outfit now? "Reevaluating" may seem unlikely to work, and let me emphasize again, it shouldn't be your first choice; it's much better if you never need to evaluate whether to check a phone notification because your phone was turned off in the first place. But sometimes, as in the example of the reduced work hours, you can't change the situation or your attention, so your only option is to try to reevaluate your thoughts.

> *In a sentence:* If you're distracted by something such as your phone, you can try reevaluating the importance of the distraction; for mind wandering, the best strategy is to talk about your situation and the desirable evaluation of the situation in the third person.

TIP 80

Test Whether You Want Social Media or Enjoy It

Suppose you really love your phone but you keep it silenced when you work. When break time comes, checking social media is a reward, of course. But even more, you find that not being able to check during the work period is a punishment; it's hard to think of anything else.

But hold on. I just said, "checking social media is a reward, of course." Did you immediately accept that as true? When you can't check your social media feed for a while, checking it may feel really urgent—you *want* to check it. When you finally do, is it pleasurable? Sure, it's nice not to feel that urgent *want* any longer. But is reading the feed actually enjoyable?

Wanting and enjoying are not the same thing. For years, brain researchers thought they had enjoyment figured out; a dopamine-rich circuit was activated every time a rat received a reward, so it seemed obvious that the circuit must relate to feelings of pleasure. More recent research shows that it actually supports the feeling of wanting and is separate from the reward circuit. Scientists thought they saw the brain saying, "Feels great!" but really it was saying, "More!"

If your brain comes to associate a particular situation or action with a reward, it will simultaneously associate it with the desire for more. The problem is that the reward may decrease—the situation or action doesn't pay off the way it once did—but the brain doesn't unlearn its "More!" response.

You may have immediately recognized your feelings about social media in the distinction between enjoying and wanting. If not, here's a little experiment you can try: For one work session, give yourself permission to check your phone as often as you want, but commit to recording three things. When you pick up your phone, rate from 1 to 7 how much you want to check it. When you're ready to return to work, record how long you were on your phone, and rate from 1 to 7 how much you enjoyed it.

The next day, allow yourself only half as many social media breaks, but make them the same average duration. Again record how much you want to check your feed as your break starts and how much you enjoyed reading it when it ends.

I'm willing to bet that on day 2 your "wanting" ratings will be higher than they were on day 1, because you had to wait longer to check your phone. But I'll also bet that your "enjoy" ratings will be no higher on day 2 than they were on day 1. What's pushing you to check social media is not the thrilling reward of seeing what people have posted or seeing how many likes you've gotten. Sure, there's some fun in that, but the main driver of your compulsion is *wanting*.

In the days following the experiment, try sticking with your regularly scheduled breaks, and **when you feel the compulsion to check your phone, try some self-talk**. Remind yourself that when you objectively recorded how much you enjoyed checking your feed, you recognized that you didn't enjoy it *that* much. The urgency you're feeling now is not the anticipation of something you'll find really fun. It's just *wanting*, a memory of what used to be a great pleasure but is now just okay. And remind yourself that you will get to that pleasure in just a little while.

In a sentence: If you feel addicted to your phone, try an experiment to test whether your habit is really something you enjoy or is just something you want.

TIP 81

Chew Gum

The evidence for this tip is not as strong as that for others, but there's some reason to think that chewing sugar-free gum might help you concentrate.

Only a handful of studies have tested the effect of chewing gum while performing real learning tasks. For example, in one study people were asked to study a twelve-page description of the human heart. In another, they were taught a strategy to multiply multidigit numbers mentally. Chewing gum (compared to not chewing it) made studying more effective and provided some evidence that people felt more alert while they studied.

In fact, the most consistent effect of chewing gum is that people say they feel like they have a little more pep. The positive effects on performance (in contrast to how you feel) aren't observed as consistently; a research review from 2011 was titled "Cognitive Advantages of Chewing Gum. Now You See Them, Now You Don't."

The upshot is that chewing gum may or may not help you concentrate; it may help some people but not others, or it may help only with certain tasks. Researchers just don't know. If it helps you, it will probably be for only a short period of time, so you might try it as an emergency boost, a bit of support that will keep you going until your next rest break. Experiment and see what you think.

> *In a sentence:* Chewing gum may help you focus your attention and stick with a task, but the research findings on this effect are mixed.

TIP 82

Fight Chronic Mind Wandering

Mind wandering has been studied for only about fifteen years, and attempts to control it are still in their infancy. Still, I can offer a few ideas that you might find helpful.

First, **don't do it on purpose!** Researchers study mind wandering in students by texting them during a lecture at random intervals (with the students' and instructor's permission), asking them to record what they were thinking about at that moment. Typically, a third of the students weren't thinking about the lecture. More surprising, about 40 percent of them had *chosen* mind wandering; they'd thought, "This is boring; I'm going to let myself think of something else." So this fix is relatively easy; if you wish you didn't let your mind wander, don't do it.

A second strategy applies to mind wandering during reading. A couple of research groups have tested whether people stay better focused on a text if they **read it aloud**. The research findings have been mixed; sometimes it helps and sometimes it doesn't, and researchers haven't sorted out whether it helps some people and not others, works only for certain types of content, or what. It's something you can try and see if you think it helps.

Two other techniques to reduce mind wandering have not been examined in experiments, but I offer them for your consideration.

When I'm involved in a task and making some progress, I find I can stay focused. The risky moment is when I finish one part of it; if I don't know what to do next, my mind is liable to drift off. My best defense is my to-do list. When I finish one task, I consult my list: What am I supposed to do next? I've described why a to-do list is helpful to planning and motivation (see tip 64). **Laying out what you're going to do may make it less likely that your mind will drift away from work.**

Finally, you might try an idea borrowed from practitioners of meditation. Some types of meditation require that you focus on one thing—for example, your heartbeat—but mind wandering can pose a problem. Some meditators set a timer to play a gentle chime every five minutes or so. It's a reminder to pull their thoughts back to their heartbeat if their mind has wandered. You might try the same technique. **Set your phone to chime every ten minutes** as a mental check-in to bring yourself back to work if you've drifted off.

In a sentence: Reducing mind wandering has been studied very little, but you can try reading aloud, using a to-do list, setting a reminder chime for every ten minutes or so, and avoiding mind wander on purpose!

TIP 83

Make Yourself Less Susceptible to Mind Wandering

I've offered ideas about fighting mind wandering during a work session. Is there anything you can do to make yourself *generally* less vulnerable to mind wandering? To change your cognitive system so you stay on task more often?

There are training regimens that promoters claim will boost your powers of concentration. Usually the training requires you to play "games" for some minutes each day. (I put *games* in quotation marks because they are not very much fun.) The games tax your ability to concentrate and mentally manipulate information. The hope is that with practice, those skills will improve.

That sounds as though it could work, but experiments show that it doesn't. People get better at the games, but they don't improve at other tasks that require concentration. At least for the time being, **there's no mental training program that reduces mind wandering**.

You *can* make your mind less susceptible to mind wandering, but the actions are kind of predictable. Mind wandering is more likely if you are hungry or sleepy, so you should **eat well and get enough sleep**. You probably expect me to say that regular exercise reduces mind wandering. Actually, there's limited evidence on that point, and the evidence we have is mixed. Exercise generally improves mood, but the relationship of mood to mind wandering is complicated and researchers aren't sure what to conclude.

The evidence is *not* ambiguous on another practice: mindfulness meditation. Mindfulness meditation can take different forms but usually involves sitting or lying quietly and paying attention to your thoughts as they occur and without judgment. Early research showed that **the minds of people who meditate regularly wander less than the minds of non-meditators do**. But of course that doesn't necessarily mean that meditation makes your mind wander less; it could be that people who already have good concentration are attracted to meditation. Further research addressed the issue by taking ordinary folks who didn't meditate, teaching them to do so, and then seeing whether their minds wandered less when performing standard laboratory tasks. They did, and more recent experiments have shown that the benefit may start as soon as a week after people start meditating.

Taking up meditation may or may not appeal to you, but surely eating well and sleeping enough do. So do at least two of those three.

In a sentence: To make yourself less susceptible to mind wandering, eat well, sleep enough, and engage in mindfulness meditation.

TIP 84

Plan Breaks, Take Breaks

You will not be surprised to learn that **rest breaks make you less susceptible to distractions and mind wandering**. You feel more refreshed and better able to concentrate after a break.

That's fine, but you might wonder if you can maximize the effectiveness of breaks. How long should they be? How often should you take one? What should you do during your break?

One answer to the first two questions has gained great popularity in the last few years. Called the *Pomodoro technique*, it entails twenty-five minutes of focused work followed by a three-to-five-minute break. After four of these sessions, you take a longer break of twenty minutes or so. There's nothing wrong with the Pomodoro technique, but **there's no research basis for the timing and duration of breaks**. So try the Pomodoro technique for a start, but don't feel as though you can't change the time values.

You can also **consider scheduling breaks by task, not by time**. I find that when I'm writing, I sometimes get on a roll and don't want to interrupt my progress with a break at a prescribed time. I'd rather work until I finish a section. That's another reason I suggest that to-do lists be composed of tasks that can be completed in thirty minutes or so (see tip 69).

Whether by time or by task, I suggest you **plan your breaks**. In other words, don't sit down figuring, "I'll work until I need a break." People who find the Pomodoro technique helpful often say something like "The first twenty minutes are usually easy, and then, when I want to quit, I can tell myself, 'You just have five more minutes until your next break!'" That self-talk is possible only if you have planned your break.

Unfortunately, research is also not much use in prescribing what you should do during your break. Experiments have compared breaks where people exercised, rested quietly, went outdoors, or worked on a different task. **There's no evidence that you're better off doing any one of these four activities than another during your break.**

I suspect that a lot of people grab their phones during a break, so it would be nice if that activity were evaluated. There are a couple of experiments showing that quiet relaxation is better than checking social media, but I think it's too early to draw strong conclusions. In addition, I can imagine that some people will be quite bothered by not being able to check social media as they work; knowing they can check it during a break helps them concentrate. (But see tip 80 on the difference between wanting and enjoying social media.)

It seems to me that a break ought to feel like a break. Do something that makes you feel revived.

In a sentence: Rest breaks help you concentrate, and there are no firm rules about their precise timing, nor what you should do during them.

TIP 85

Regroup or Move Along

Not long ago I was trying to think of an opening for a talk I was writing about technology and reading. I couldn't think of anything, so I started aimlessly casting around Google, hoping for inspiration. Predictably, I found nothing useful and started reading stuff unrelated to work. Then I got mad, told myself, "I need to *think*," and two minutes later my mind was wandering.

What should I have done? An obvious answer is "Take a break," but suppose I had just taken a break?

Another choice is to **regroup**: evaluate the task I've undertaken and the methods I'm using. Why am I making no progress? What am I trying to do? What have I tried? What's gone wrong? Maybe I don't need a clever opening for a talk on reading and technology; after all, educators are already interested in the topic. Or maybe I should have an opening but I should ignore Google and reflect on my own digital reading habits or those of my kids.

If regrouping doesn't work, consider **moving along**. I could work on the rest of the talk and return later to the nut I can't crack. Perhaps the fresh perspective will bring fresh ideas.

The important point here is that you must monitor your own distraction and notice when you seem to be stuck in a revolving door on a particular problem. Then you must avoid the stubbornness that often comes with a lack of progress. You think, "I can't quit now—I haven't solved it!" But you shouldn't throw good time after bad. Rethink your approach.

In a sentence: We are especially susceptible to distraction when we work on a problem without making progress, so when you feel that happening, you can either regroup by trying a fresh approach to the problem or set the problem aside temporarily and work on something else.

For Instructors

Telling students about the strategies I've listed is easy enough, and they are not especially difficult to implement. The biggest problem lies in persuading them that they are needed or helpful. I've cited data showing that students do not think that multitasking carries a cost. You might consider a demonstration showing that it does. Here's one idea.

Find two brief videos (lasting, say, five minutes each) that relate to course content. For each, write six questions that test students' comprehension.

Show the first video to your students and administer the brief test. As they watch the second video, periodically say "Ping!" and hold up a poster board with a question written on it. This is a simulated text message. Students should read the message and write a response on a piece of paper. The messages should be simple questions of the sort they might get via text but should appear only briefly; if they don't answer quickly, their friend will feel ignored.

Base three questions on video content that appeared at the same time as a "text message" and three on content that appeared at times when students weren't distracted. Have the students compare their comprehension when they were multitasking versus when they were not. The goal is to show them that doing two things at once is actually harder than they think.

The other principle that students probably don't grasp is that distraction happens in stages, and as you move through each stage, it gets more difficult to turn your attention back to your work. To help them understand, try this demonstration. Ask half of your students to turn their phone off for one hour in the evening. At the end of the hour, ask them to rate how hard it was to have their phone off, from 1 (easy) to 7 (terrible). The other students leave their phone on and nearby but don't touch it for sixty minutes. They, too, should rate their discomfort at the end of the hour. The following evening, everyone switches tasks.

Most students find that ignoring notifications is harder than having the phone off because every ping is a reminder that they can't check their phone. The goal is to drive home the message that a little advance care in shaping their environment will make it easier for them to maintain their focus.

Summary for Instructors

- Tell your students about the strategies described in this chapter.
- Use demonstrations to convince students that they cannot multitask and that distraction occurs in stages.

CHAPTER 13

How to Gain Self-Confidence as a Learner

A common movie plot features a teacher helping "bad kids" show the world—and themselves—that they are actually quite smart. Movies such as *Lean on Me*, *Freedom Writers*, and *Stand and Deliver* depend on viewers accepting that people can be grossly misinformed about their academic ability—that they can be smart yet not know it. And most viewers accept that premise; it makes sense that factors other than the characters' actual smarts contribute to their self-confidence—factors like whether their parents encourage them, for example.

Yet most people don't believe that the same could be true of themselves: "Someone else might not understand herself very well, but my case is simple: I felt dumb in school because I *was* dumb. It's true that, like the kids in the movies, my parents never encouraged me, but I didn't deserve encouragement, because I got bad grades."

A lack of self-confidence matters, because it affects your academic success. For one, it shapes **how you interpret setbacks**. When a college student who thinks of himself as a good learner flunks an exam, he assumes that he didn't study hard enough and that he can do better next time. A student who was never sure he belonged in college in the first place might take the failing grade as evidence that he doesn't.

Your self-confidence also **affects your aspirations**. For example, someone who always dreamed of being a registered nurse but sees herself as a poor student may conclude that she could never get through nursing school and so choose another career.

Your self-confidence as a learner comes from **your academic self-image**: Do you see yourself as someone who learns easily or someone who struggles? Unsurprisingly, your self-image is shaped, in part, by grades and other feedback you've received over the years, but three other factors matter, too: who your friends are, who you compare yourself to, and the family values you grew up with.

There are no simple rules about how these four factors combine and therefore how to change your self-image if you think it ought to change. Still, if you lack self-confidence, it will benefit you to realize that at least part of that feeling comes from factors other than your competence and you should have more self-confidence than you do.

WHEN THINKING ABOUT YOUR SELF-CONFIDENCE AS A LEARNER

What your brain will do: It will construct an academic self-image based partly on your prior learning success but also on relationships, who you compare yourself to, and your values; this self-image determines your self-confidence.

How to outsmart your brain: Take steps to change your academic self-image once you know the factors that contribute to it.

To get started, I'll ask you to reflect a bit on the four contributors to academic self-image:

Feedback: What kinds of messages have you gotten about your competence from the world around you? Do you generally succeed when you try to learn? Did your teachers think that you belonged in advanced or remedial classes? When you had an academic setback, did your parents tell you that you could do better next time if you studied harder, or did they seem to assume that you weren't cut out for school?

Social relationships: Your views of other people develop as you watch their behavior; so, too, your view of yourself is influenced as you observe your own behavior, including the people you associate with. Do your friends see learning as an important part of their lives? Do they make time to learn new things?

Comparisons: A student who gets mostly Bs may think of herself as capable if she compares herself to her best friend, who gets mostly Cs. Or she may think of herself as "the dumb one" if she compares herself to her sister, who gets mostly As. Did your parents or teachers compare you to other kids, and did you agree with those comparisons?

Values: In a family that prizes education, a child is less likely to question whether he really belongs in school because his parents so firmly assume that he does. Other parents believe that there are many paths to a good life and learning may play a larger or smaller role for each person. Would you describe learning as a family value when you were growing up, and did you embrace that value or rebel against it?

In this chapter I'll offer ideas for examining and perhaps rethinking the four factors that contribute to your academic self-image. The purpose is for you to come to a clearer sense of yourself as a learner and to be sure that your self-image and the self-confidence that follows are realistic.

TIP 86

Rethink What It Means to Be a Learner

Most people develop their idea of a "good learner" at school. The good learners are the ones who are not corrected when they read aloud. They raise their hands to answer teachers' questions, and they never seem confused by math. Teachers may not overtly label them "the smart kids," but they don't have to. It's obvious to everyone.

The early elementary school years are formative, so once you develop this concept of a good learner, it's hard to shake. But it's limited in two ways.

First, it puts a premium on speed. School curricula are packed, so teachers feel pressured to keep the pace brisk. The students who catch on to ideas quickly are at an advantage. The plodders might arrive at the same, or even deeper, understandings, but they might never have the chance to show their intelligence.

Second, this description of a good learner makes it a characteristic intrinsic to the person, something they just *are*, like brown-eyed or sixty-eight inches tall. But as you've seen in this book, **learning is effective because of what you *do*, not who you are**. If you've had trouble learning in the past, it's not because you're not a learner. Maybe you're slower than others, but anyone who does the right things to learn is a learner. It's really part of your birthright as a human.

Does that seem a little overstated? **Think about what you've learned**

outside of school. Maybe you learned to play a sport, to excel at a video game, to negotiate complicated social relationships among your friends, to play an instrument, to deal with a difficult parent, or to navigate a tough neighborhood. There's probably something you're pretty good at, but even if you're not great at anything, you've still done a lot of learning. Maybe you did most of your learning in informal settings. If you're reading this book, you're thinking of changing that, but that's not as big a change as you think, especially because you're now armed with the strategies you've learned here.

And once you're no longer a student, **the measure of "success" will be different**, so you shouldn't assume that your experience will be the same as it was in school. Outside school, successful learning is often coupled with other abilities or skills. For example, suppose you're a sales representative and you've been using some new project management software for the last six months. It's worked well, so your boss wants the engineers to use it, and she asks you to persuade them. That task certainly requires learning—you need to get up to speed on how engineers think about projects—but it's as much an interpersonal task as a learning task.

School prizes raw learning, but the workforce prizes many other skills: building trust with coworkers, for example, or having the courage to try something new. Keeping that in mind may be the most important way to modify your thinking about what makes a "good learner." Once you're outside school, **you don't need to be *great* at learning; you need to be okay at it but also achieve proficiency in other skills.**

It's a point made by Scott Adams, the cartoonist behind *Dilbert*. He wrote that one path to success is to become extremely good at one thing, but of course it's difficult to become *that good* at anything. It's much easier to become pretty good at two or more things. He noted that he can draw, but he's no artist. He's funnier than most people but not as funny as professional stand-up comedians. He also has a business background, so that's three areas of competence, which very few people have simultaneously, and the result is an extremely successful comic strip set in an office.

If you're thinking, "I've never been a good learner," ask yourself whether you really need to be great at learning or whether being "pretty good" at learning, combined with some other skills, will yield a first-rate combination. If you use the strategies in this book, being "pretty good" at learning will surely be within your reach.

> *In a sentence:* Remember that learning is something you do, not something you are, and that the definition of *successful learning* changes once you're out of school; you need to be good at several things, not excellent at one.

TIP 87

Be Around Other Learners

To a greater degree than many of us like to admit, our behavior is influenced by the people around us. Evolution has left us with a mind that is sensitive to what others do and primed to mimic them, because if everyone else is doing it, it's probably the safe and smart thing to do. That's why people laugh more when a television show has a laugh track. People go to restaurants that they see are usually crowded and avoid ones that are usually empty.

When the decision is more consequential than "Should I laugh?" or "Where shall we eat?" we care less if strangers are doing it and are more influenced by close friends and family. For example, most people wouldn't pay for a new game just because they hear it's popular, but if a few friends say they like it, that may be enough.

If you mimic what your friends and family do, you're guaranteed social support when *you* do it. For example, if most of your friends take learning seriously, they will make it easier to go to the library on the nights you don't feel like going. They will listen and commiserate when your learning isn't going well, and they'll cheer when it is. They're in a good position to provide practical help with your work, like offering studying tips they've found useful.

It's not that friends who aren't interested in learning are bad friends, it's just that the social support for learning in particular doesn't come as naturally to them. They won't cajole you into studying when you don't feel like it, because they themselves aren't studying. They'll sympathize when your work isn't going well, but it doesn't have quite the same feeling because you know they probably don't experience it the way you do. **If you care about learning and the people in your social group don't, there's a part of your life in which you feel a little lonely.** We like to affiliate with people like ourselves.

I've actually met people who *hid* their interest in learning because they were afraid that their friends would reject them. A few years ago I received a poignant email from a high school English teacher about one of her students. He was very devoted to the football team, and he was also devoted to reading literature, but he was so sure that he would suffer socially if that news got out, he wouldn't even broach the subject with his friends. He ached to discuss books with someone and so asked the teacher if he could sometimes talk to her after school.

Obviously, **you shouldn't drop friends who aren't interested in learning, but you might add some who are.** Whether you're reading up on science for your own pleasure, seeking high grades in hopes of medical school admission, or trying to read more weighty news sources to understand contemporary politics, being around people who share your interest will offer the social support that humans crave.

In a sentence: We are social beings, and we are influenced by what our friends and family do; being around at least a few other people who care about learning will make it easier for you to express that side of yourself.

TIP 88

Compare Yourself to Yourself

Which activities or attitudes have the greatest influence on your self-image? It seems as though they should be the ones you feel are most important to you, or maybe the ones that you spend a lot of time on. But on reflection you'll realize that's not right. A teenager who loves video games and spends a couple of hours each day playing may not see himself as a gamer. Why? Because everyone he knows plays games just as much. But if none of his friends reads, they'll think of him as "the reader" in their group even if he reads just two or three books each year. It's the contrast that matters.

The comparisons that affect your self-image are not just the ones that your friends make; **you pick people to compare yourself to.** Your self-image can vary widely depending on your selections, and there's no good way to know which comparisons make sense. Sometimes we make comparisons to flatter or reassure ourselves. It's a tragic truism that people with a substance abuse problem look for someone further out on the limb: "I may drink a lot, but I'm not as bad as *him*."

But we don't always make comparisons that gratify us. A friend told me about a graduate student in his lab who feared he was going to flunk out because he couldn't hack statistics. Actually, he was near the top of his class, but he compared himself to his husband, who was getting a PhD in data science.

Anybody can observe these situations from the outside and say, "Your comparison doesn't make any sense, and it's distorting your self-image." But how are you supposed to know who is a good comparator?

The question brings to mind a nineteenth-century Chasidic teaching, which I'll freely adapt. Everyone should have two pockets. In one, keep a slip of paper on which is written: "You are the crown of God's creation, closest to the angels." Reach into that pocket when you feel sad and worthless. But if you feel too high-and-mighty, reach into the other pocket. There, you keep a slip of paper on which is written: "God created the earthworm before you."

There's always someone you think is ahead of you and always someone you think is behind, and I can see the advantage of using that fact to manage my emotions. But I don't trust myself to use it wisely. I'm exactly the kind of guy who, when I'm feeling low, would reach for the earthworm paper.

Rather than shooting for clever comparisons, **compare yourself to yourself**. That means tracking your goals and your progress in meeting them. I've already suggested that you do this (see tip 65), so I'm not proposing any extra work; rather, it's an extra use to which you can put your recorded goals. In reflective moments, most people agree that comparing ourselves to others is at best unproductive and at worst damaging. What matters is striving to be the best we can be, and what other people are doing or not doing is irrelevant. Remind yourself of this when you start wondering if you're measuring up to your peers. Take out the computer file or journal where you record your goals, and review your progress.

In a sentence: It's natural to compare yourself to others, and comparisons contribute to your self-image, but they are seldom helpful; compare your present self to your past self when evaluating your progress.

TIP 89

If You Didn't Get Practical Learning Advice from Your Family, Get It from Others

Even if parents seldom discuss family values in so many words, children know what their parents care about via silent messages conveyed by actions. Children observe what their parents spend money on, what they devote their time to, who they think deserves respect, and what's important enough to merit a household rule. These clues make clear the value that parents place on religious observance, social advancement, a political perspective, financial success, learning, and more.

Children raised in families that value learning tend to do well in school. They take more challenging courses, earn higher grades, and are more likely to graduate from high school and continue on to college. That's partly because parents who think of learning as a family value tend to have more money and more education themselves, and so they can more easily give their children advantages; they can hire a tutor if need be, for example. But in addition, their children enjoy a deep confidence that they belong in school and can succeed.

In contrast, some people grow up with parents who are uninterested in learning. Other people's parents are interested in it but lack the time and money to act on that value. Either situation can lead to a long-term, nagging feeling that you just don't fit in at school.

In my years in higher education I've met scores of students who felt this way, but the most memorable example was one of my first graduate students. He was getting great feedback on his work, yet he was plagued by uncertainty, a feeling that he was missing something. He thought that there was a set of unwritten rules about how to act in graduate school and he was the only one who didn't know them, because of his background— he was the first in his family to attend college.

That feeling may have been a carryover from high school and college, and there his suspicion would have made sense. **Parents who themselves felt comfortable at school often have some knowledge about how to succeed there.** They provide advice to their kids and advocate for them. For example, if you fail your first college exam, your father might tell you that he did the same but was able to bounce back. Or maybe your mother advises you to see the professor and ask how to succeed next time. Most important, if your parents have always assumed that you would graduate from college, you will feel you belong there, and one setback won't make you question that.

What should you do if your parents don't have that knowledge? In high school, **your teachers can help**. Pick your favorite teacher, even if it's been a few years since you were in his or her class, and ask for the guidance you need, even if you're unsure of exactly what you need help with. Talk it out. Most teachers will not see this request as a nuisance. On the contrary, they will be pleased that you sought them out.

College has a different set of rules for success than high school does. Your brain doesn't change, so studying and learning are the same, but the organization of the school is different, so you're faced with unfamiliar problems. How should you pick a major? A spot just opened for a great course, but you're three weeks into the semester; if you sign up for it, will you be able to catch up?

Unfortunately, many colleges ask faculty to advise students on these matters, and professors often don't have the knowledge or motivation to do the job well. If your advisor is not helpful, try the director of the undergraduate program for your major. (That person might also be able to assign you to a different advisor. Ask.) Or try the Office of Student Affairs or the Office of the Dean of Students or whatever it's called at your school. **Every school has an administrative arm designed to help students understand the system.** You won't be bothering anyone—helping you with this sort of thing is their job, and the job exists exactly because the system is confusing.

> *In a sentence:* Some children gained self-confidence as learners and practical advice about school from their parents, but people who didn't can gain those things from other sources.

For Instructors

How can teachers contribute to making everyone feel capable and eager to take on challenges?

What you'd like is for self-doubting students to succeed in some way so that after you celebrate their accomplishment, you can gently push them to acknowledge their success. You want, in essence, to say, "See? You thought you couldn't do it, but you can."

But you may wait a long while for a student to feel that he has succeeded. And you'd rather that students focused on processes anyway. That is, you'd like for them to feel pride in working hard to prepare for an exam—in preparing a really complete study guide, for example—

even if their performance on the exam was only so-so. That's asking a lot, but I think it's worth articulating; even if students don't embrace the message, they may register it and understand its importance.

That's a natural segue to another way that your feedback can support students' self-image: help them identify which tasks give them trouble. They may think, "I'm a bad student," but as we've seen, many steps go into academic tasks like test preparation. If they know what they do well and what they don't, it might change their self-image: "I'm not a bad student, but I need to take better notes."

Personal connection may be more than a facilitator of these methods; it can be a potent, if indirect, source of positive feedback to students. Some studies of community college students show that a personal connection with someone at the school can be a powerful motivator for a student who feels hesitant about his or her place in the school. These studies show that the connection did not need to be made with a faculty member; sometimes it was with a cafeteria worker or someone on the secretarial staff. What's more, the person often did not serve any traditional mentoring function. The relationship might have been meaningful to a student in two ways. First, on a large, anonymous campus, it was someone who would notice if he didn't show up and would even feel disappointed. Second, it was someone who, as an employee, had seen many students at the college and seemed to take for granted that the student belonged there, which served as a silent affirmation of her status.

Helping students feel at home in school and confident about their learning is one of the most challenging tasks teachers face, partly because students' self-confidence is determined in some measure by factors outside of school and also because the messages that teachers send students that affect their self-confidence can be extremely subtle. Still, monitoring the messages we send is worth our care and attention, as they can have a profound impact on students' long-term success.

Summary for Instructors

- Prompt students to acknowledge their successes.
- Help students feel good about engaging the right processes for academic work and see that as progress, even if their grades aren't great.
- Help students identify which parts of academic tasks they do well and which make them struggle; then you can help them troubleshoot the hard parts.
- Forge personal connections. They go a long way in making students feel comfortable and confident in school.

CHAPTER 14

How to Cope with Anxiety

Some anxiety is not just normal, it's helpful. Anxiety prepares you for action by mobilizing your body to either flee or fight. What's more, it sometimes informs you. You might observe your body's reaction—a pounding heart, for example—before you're fully aware of what the threat is. Anxiety tells you that there's a problem so you can scan the environment to learn more about it.

When you think of anxiety and learning, you probably think first of test anxiety, of someone who knows the content well but fails to show it on a test because of nervousness. As I said in chapter 8, it's pretty typical to feel some anxiety while taking an exam. What's less common is for the anxiety to feel overwhelming and to affect you not just at test time but when you're performing other learning tasks, like reading or taking notes.

Anxiety goes from "helpful" to "damaging" when you habitually spend time and mental energy checking the environment for threats that aren't there. The spider-phobic thoroughly scans any room before entering to be sure the coast is clear and then keeps scanning it once he's there. That consumes attention and makes it hard to hold a conversation or even think. And anxiety can affect behavior as well as thinking. The spider-phobic might refuse to enter his own living room because he has seen spiders there before.

Where does this maladaptive anxiety come from?

There's little doubt that **a moderate proportion—perhaps a third—can be assigned to our genes**. That doesn't mean your DNA determines "Thou shalt be anxious" as inevitably as your eyes are destined to have a particular color. It means that you have a predisposition to the kind of vigilance, the watchfulness that easily blooms into anxiety. But what prompts it to grow?

There are two theories. One suggests that **anxiety is a product of the same type of learning observed with Pavlov's dog**. You ring a bell, then feed the dog. Repeat that enough times, and the dog expects to be fed when it hears the bell and therefore salivates.

The same process can make you anxious about learning. I'll use math as an example. Suppose that during a class you are asked to solve a math problem at the blackboard. You can't solve it and you feel humiliated. Repeat that a few times, and you expect to feel humiliated every time you are asked to go to the board to solve a math problem, just as a dog expects to be fed when it hears the bell. The anticipation of humiliation makes you anxious.

But it doesn't end there.

You know that math class is where you might be asked to go to the board to solve a problem, so now you get butterflies in your stomach the moment you walk into math class. And working math problems at home reminds you of working them at the board, so you feel uneasy when you do that. Anything associated with math can become a source of anxiety. This theory of anxiety emphasizes the way that something that started as neutral (math) becomes associated with something negative (frustration and shame).

Another theory helps us understand how **anxiety can get out of**

control. The feeling of anxiety is so unpleasant that you always have your feelers out, so to speak, monitoring the environment for the thing you find threatening. This monitoring process is unconscious, but what's *not* unconscious is the feeling of nervousness, of anticipating that you might encounter the thing you dread. So you think, "Things must be really bad, because I'm very nervous *and I can't find the thing that's making me nervous*." Such thoughts make you even more concerned about threats, so you look even harder for them, you don't find them even though you think they must be there, and the vicious cycle continues.

Now, you may have noticed that there's actually something rational about what we're saying is irrational anxiety. In my example, math anxiety began with difficulty working problems at the blackboard. Shouldn't we just say, "Being bad at math makes you anxious about doing math?" Research indicates that that's a factor, but it can't be the whole explanation. A subset of the people who have math anxiety are actually pretty good at math. And there are others who are terrible at math but don't feel anxious about it. How can that be?

It appears that **a person's interpretation of events is crucial**. You're much more likely to feel anxious about math if you think a failed test tells you something important and unchangeable about yourself. If math is unimportant to you, a bad test score doesn't make you anxious. You're also okay if you do care about math (so you're upset about your low test score) but you think you can improve if you work harder. You get anxious only if you care and feel helpless.

When we turn our attention to reducing anxiety, two things become clear. First, given that your interpretation of events matters more than what actually happens, it would seem that **the main thing we need to do is give you a better way to think about what happens**. Second, **we shouldn't expect anxiety to go away quickly**. Even with a better way to think about events, people need to unlearn their old associations and ways of thinking. It's like any other difficult task; you wouldn't expect to run a marathon your first day at the track. You need to work at it and expect modest progress.

In fact, eliminating anxiety takes long enough that most psychologists would say that it shouldn't be your goal. If you get anxious when you take a test, offer an idea in class, or work on a project with people you don't know, the important thing is to be able to take the test, offer the idea, or work on the project. Your goal should be the management of your anxiety. Don't feel "I can't do that task until I no longer feel anxious about it." Your target is to be able to do it despite your anxiety.

IF YOU SUFFER FROM ANXIETY

What your brain will do: It will scan your surroundings for threats and continue to do so even if nothing threatening is observed. This scanning will heighten your anxiety in an upward spiral and will occupy your mind, making it difficult to focus on learning.

How to outsmart your brain: Focus on reinterpreting your thoughts to manage your anxiety.

This chapter includes a range of strategies. All have been shown to be effective in scientific experiments, but that doesn't mean that each one works equally well for every individual. I encourage you to try different strategies and see what works for you. It's probable that a single action won't do the trick and you'll need to put multiple strategies into play. Be patient. This will take time and practice.

TIP 90

Evaluate Progress as Any Improvement in Doing What You Want to Do

Feedback is important to the work of managing your anxiety. Because different strategies are more or less effective for different people, you must be able to get a sense of whether a specific tip works for you. How do you know if things are improving? You probably thought the definition of *success* would be "feel less anxiety," but I've already "noped" that idea.

Consistent with your goal of managing anxiety, not eliminating it, **define *success* as doing what you want to do, even if it makes you anxious.**

At this point you may be snorting "Some tip! 'Ignore your terror and just do it.'" Well, yes. **Feeling anxious is uncomfortable, but it's not dangerous.** That's hard to remember when your heart is pounding and your palms are sweaty—your body is telling you quite clearly, "There's a problem here!" But in calmer moments you know that everything is actually fine and you can't be harmed. You can push through. You may be very uncomfortable, but you're not in danger.

Some years ago I had a student who exemplified this idea in a way that inspired me. She showed minimal evidence of social anxiety in other interactions, but every time she spoke up in class, a flush would start on her chest and rise to her neck. That and her somewhat halting speech showed that speaking in a group made her extremely anxious. But speak she did.

I'm sure I was seeing the product of a lot of work. She was articulating complicated ideas that required her to speak for sixty seconds or more, and I'll bet she started with brief comments. Maybe she even planned actions in stages, for example:

- Say something short once a week in class.
- Say something short in each class.
- Describe a more complete idea (taking, say, a minute) once a week.
- Give a short presentation in class.

I urge you to do the same. **Count as a success doing a little of what you want to do.** Perhaps all you say is "I just want to add that I really agree with that point" in support of someone else's comment. If you avoid leaving the house because you fear social interaction, maybe the first step would be to take a walk around the block and promise yourself to say "Hello" to a passerby. The next tip elaborates on why the goals you set should be small.

Here's what you should *not* be tallying in your mind: how you compare to others or how you compare now to where you'd like to be. Such comparisons are an invitation to unfair self-criticism and deciding that you're weird or a loser. The right comparison is where you are now and where you used to be. That's your focus. That and the next small step you can take.

In a sentence: You should evaluate whether the strategies you use are working, and the right definition of *working* is not that you're feeling less anxious but that you are making some progress in doing what you want to do.

TIP 91

Avoid These Four Common Responses to Anxiety

Anxiety comes with a set of common thought patterns. Unfortunately, they don't improve things and, in fact, make anxiety worse. Here I'll list better alternatives to four common responses.

Don't give up. Don't fail to do things because you're anxious. For example, don't tell yourself, "I'm too anxious to talk to my advisor" or "I shouldn't take the advanced course even though I qualified, because the thought of it makes me uncomfortable." You're not required to seek out situations that make you anxious, but you still need to do the things you need to do. And you can. Anxiety makes you uncomfortable, not incompetent.

Instead, review your past successes. Remind yourself, "I have done this sort of thing before. Parts of it were hard for me and I did feel uncomfortable, but I got through it. I can do it again."

Don't catastrophize. When we're anxious, our thoughts easily run away from us—we predict that things will end badly and there will be lasting consequences. So you don't just think, "There's a chance this presentation won't be very good"; you think, "My presentation will be terrible, I'll flunk the course, and I'll never be a radiation therapist."

Instead, think from a distance. Try to make your assessment more rational by depersonalizing it, thinking about the situation as though it were happening to someone else. In other words, think, "Take someone like me, who's a pretty solid B student. Now suppose that person gives a really terrible presentation on his project. The presentation is worth ten percent of the grade. Is that person likely to fail the course? What's more likely to happen?"

Don't deny that you're anxious. Don't just keep repeating to yourself, "Don't be anxious don't be anxious don't be anxious." Don't think, "I can't be anxious about *this*. This is nothing. Only a loser would be anxious about this. Okay, I am just *not* going to be anxious about this." Suppression is not a winning strategy in the long run. You can't keep anxiety out forever.

Instead, use suppression in the short term. Denial and suppression should not be your long-term plan, but suppression might be useful in the short term, especially if you have a plan to deal with the underlying issue later. For example, you might say to yourself, "I'm feeling nervous about that test I have on Friday, but I'm with friends now, and it's okay for me to have fun. I have time mapped out each night for studying, and I can think about the test then. I've scheduled plenty of time for that, so it's okay not to think about it now."

Don't self-medicate. Alcohol and other drugs may provide temporary relief from anxiety, and under a doctor's supervision the limited use of medication may make sense for you. I've suffered from anxiety myself, and when it got really bad, I was too frazzled to put into practice any of the tips outlined here. Medication made space in my head for me to address my anxiety. But using drugs or alcohol solely for temporary relief from anxiety and doing so without consulting a medical professional is not a road to improvement.

In a sentence: Anxiety is usually accompanied by certain thought patterns that make it worse, so it's useful to be able to recognize them and direct your thoughts away from them should they occur.

TIP 92

Reinterpret What Your Mind Is Telling You

I've described (and you may have experienced) how anxious thoughts can spiral out of control: your anxiety prompts you to search the environment for a threat, you find none, and that makes you even more anxious. How can you interrupt that cycle?

Here's a three-step process to slow down your runaway mind. I urge you to write down your thoughts as you go through the first two steps. Writing helps because choosing which ideas are worth recording forces you to weigh and evaluate them.

To begin with, **normalize** your thoughts, rather than fighting them or addressing them directly. "This is normal, this thing that's happening to me. It sucks, but it's normal. I'm not crazy or weak, any more than someone who gets migraine headaches is crazy or weak. And it's not unacceptable for me to feel anxious. It's just something that happens to some people."

When you've been over that bit in your mind, it's time to **evaluate**. What are the chances that one of the things you're contemplating will actually happen? And what would the consequences be if it did? Is it really likely that the teacher will call on you when you don't know an answer? Does that happen routinely? Or do you routinely worry about it, although it almost never does? Negative thoughts can seem powerful, but your thoughts can't cause anything to happen. Thoughts are unsubstantial, temporary, and, we might add, private.

Suppose the terrible thing you contemplate does happen: you're called on and you don't know the answer, you fail a test, or the other members of your study group think you're poorly prepared. Well, then what? If you fail a test or even fail a course, your future is not down the drain. If you let down your study group, you apologize and try to make it up to them next time.

The final step is to **reengage**. You've normalized your anxious thoughts, you've evaluated them, and now it's time to get out of your head, beyond your thoughts. You need to reengage with the world. You need to show yourself that the thing that prompted your anxiety hasn't beaten you. It can be a baby step. Maybe you write one paragraph of the paper you're working on. Maybe you decide that you won't press yourself to say anything at the next meeting of your seminar, but you won't hide; you will make eye contact with people who are speaking and nod if you agree with what they're saying.

If you have something to do that you know will make you anxious, it is a good idea to **use this three-step process a day or two beforehand**. If you wait until you're feeling panicky because of a class presentation, you will be too jittery to put any of this thought work into practice. Instead, try to normalize, evaluate, and reengage a day or two before the presentation. As you're getting tense about the impending event and your thoughts start to spiral, you will be able to say to yourself, "I went through all this the other day. I figured out that this presentation is not as big a deal as I was making it out to be."

This is hard work. It's easy to say, "Normalize your thoughts," but much harder to do it. In fact when you start, it may feel nearly impossible—but **it gets easier**. And remember, all forward movement is progress.

In a sentence: Use a three-step process—normalize, evaluate, and reengage—to reinterpret what your mind is telling you when you're anxious.

TIP 93

Reinterpret What Your Body Is Telling You

Anxiety involves both your mind and your body, and your anxious body complicates your efforts to calm your runaway mind. You may experience **a hammering heart, tense muscles, sweating, dizziness**, or some combination of these. It's hard *not* to interpret these feelings as an indication that there's danger in the environment. You know quite well that you're feeling a fight-or-flight response.

But there's actually another way to think about your body's reaction: **you get the same feeling when you're excited.** Your heart would pound if you watched your best friend surprise her girlfriend with a marriage proposal, if your cousin were nominated for an Oscar, or if your favorite team had the chance to beat its rival with a last-second field goal.

I do a good deal of public speaking, and my heart pounds before every talk, but not due to anxiety. It's excitement. And a bit of excitement (or, as it's more typically called, arousal) helps you do a better job. If your arousal isn't high enough, you're sleepy. Next time your heart pounds and you start to sweat, don't start talking to yourself about how anxious you are. **Think of yourself as excited.** Your body is telling you that it's ready for adventure!

In a sentence: Don't assume that certain physical symptoms necessarily mean that you're anxious, because you feel the same symptoms when you are excited.

TIP 94

Tame Your Wild Thoughts with Mindfulness Meditation

I've said that the anxious thoughts ricocheting around your mind are uncomfortable, even scary, but it's important to remember that they don't *make* bad things happen. They don't, on their own, have any power. That's easy to say but much harder to believe.

Mindfulness meditation can help you change your relationship with your thoughts. **Mindfulness meditation is simply the practice of observing your thoughts, feelings, and sensations, and doing so without judging them and without criticizing yourself.** It's not "thinking about nothing"; it is being in the moment.

People who are more qualified than I am have generously made detailed instructional resources freely available on the internet. But here's a quick overview to give you the idea. You set a timer for as little as two minutes (for starters), sit (or lie) comfortably, and breathe slowly. In many varieties of meditation you focus on your breath or your heartbeat. Your thoughts zip about, some bouncing around, some streaking past. You simply watch them go, refraining from judging them or judging yourself for *having* these thoughts, and return your attention to your breath. Practitioners often use imagery to help let go of thoughts. You imagine them as leaves passing by on a stream, as clouds drifting by on the wind, or as waves crashing on a beach. Each thought comes, it recedes, and then it's gone. That's it.

"That's it," but people who have meditated daily for years will tell you that (1) it's hard work and (2) they are still learning new things in their practice. Yet even a beginner can see benefits. That's why medical practitioners have suggested mindfulness meditation to patients with a broad range of ills both mental and physical. One of the most common is stress and anxiety, and researchers report positive effects of even brief mindfulness meditation training. Indeed, hundreds of medical centers around the United States (including the University of Virginia, where I work) have mindfulness-based stress reduction (MBSR) programs.

Why does watching your thoughts reduce anxiety? Two mechanisms may be at work. One is that **you come to know a feeling of quiet in your mind**—what it's like *not* to have a torrent of troubling thoughts. Having felt that mental quiet frequently makes you more confident that you can find it again when you're waiting for a final examination to begin or are in some other situation that makes you anxious.

Mindfulness meditation might also help you **improve your ability to recognize your thoughts more fully** rather than react to them emotionally based on a glimpse of them. Thus, when you sit alone at a restaurant table, waiting for your friend who is fifteen minutes late and has not responded to your text, your first reaction may be mounting anxiety that something terrible has happened. But some introspection leads you to realize that your anxiety is really fueled by worry that your friend has simply decided not to come. And that thought is somehow easier to reject as irrational. Your old friend wouldn't suddenly ghost you.

Mindfulness meditation sounds daunting, but it fits the "baby steps" approach quite well. No one needs to know you're doing it—as I noted, there are lots of **tutorials on the internet** and plenty of **apps** (e.g., Headspace, The Mindfulness App, Calm) to guide you. You can start by meditating just two minutes each day—consistency from day to day is more important than the length of each daily session. If you decide to give meditation a go, do keep in mind that initially you will "fail" a lot—that is, you will find it hard to focus as you're supposed to. Meditation is a skill like any other, and it gets easier with practice.

There's no guarantee that mindfulness meditation will be a good fit for you, but it's low cost to try, and it makes a world of difference for some.

In a sentence: Mindfulness meditation is easy to try and is a great help to some people in dealing with anxiety.

For Instructors

On average, 20 percent of the students you teach are anxious. Schools typically have policies dictating accommodations for students who have had a formal diagnosis of anxiety. What about those without a diagnosis?

I make a general appeal to students to self-identify, saying something like "If you have any health issues—for example, if you're battling anxiety or depression—please email me or stop by my office so we can work together to be sure you get the most out of this class."

I always start by asking the student what he or she would like me to do. Part of the reason is that students know better than I do what they find troubling, and part is that I want them to take responsibility for addressing the issue themselves, rather than my leaping forward with suggested remedies.

My rule of thumb is that I won't provide an accommodation to a student with anxiety that I wouldn't provide to a student without anxiety. For example, I won't let an anxious student miss class, submit work late, or simply not participate in group work. That may seem harsh, but it's in keeping with the approach I've emphasized throughout this chapter: you don't simply not do things because they make you anxious. Anxiety is not a disability, and students can do everything that's expected in the course.

Examples of accommodations I would make:

For test anxiety: Sitting in a particular seat. Wearing a hood during a test. Taking a sixty-second walk during an exam.

For anxiety about class participation: Posing a question and giving students two or three minutes to write out their answers, then calling on the anxious student to read her response so that she need not compose an answer on the fly. Adopting a classwide policy of using name tents that a student sets on end to signify that he wants a conversational turn, thus making entry into the discussion easier. Encouraging very brief comments in the conversation.

For generalized anxiety: Offering help in thinking through how to break down large assignments into smaller tasks. Providing clearly stated, written explanations about what's expected for assignments.

Don't think that you need to "treat" or resolve any student's anxiety. You're not trained for that, and anyway the student isn't looking to you for that help. He or she just wants to succeed in your class.

Summary for Instructors

- Follow your institution's guidelines for accommodating students with a diagnosis of anxiety.

- Ask students who are anxious or struggling to identify themselves to you so you will be aware of the reasons they are struggling.

- Hold anxious students responsible for all the work in a class (again, subject to any guidelines set by your school).

- Offer the same accommodations that you would offer any student.

- Remember that you're not responsible for treating or resolving your students' anxiety.

CONCLUSION

In the fall of my third year as a professor at the University of Virginia, I was required to provide a written account of what I'd achieved in research and teaching to that point. Two senior professors were to read this review and meet with me to offer guidance on how I could improve. I was eager for their feedback, as I was three years away from a much more serious review, one that would result in one of two outcomes: I'd be promoted or fired.

But I got no feedback during that meeting because my so-called guides did a very professor-y thing; they got into a debate with one another and ignored me. Professor X said that my work looked promising, but she had noticed that there was no sense of *fun* in the documents I had prepared. She felt that all great scientists regard their work with a certain playfulness, and I seemed awfully solemn. Professor Y quickly disagreed, saying, "Fun? Going to parties is fun. Research seems serious because it *is* serious." For fifteen minutes they argued about whether or not critical thinking is fun. Then they remembered why we were all there. Each told me, "It looks like you're doing fine," and the meeting ended.

That debate came to mind as I was finishing this book, because I've never even hinted that learning might be fun. In fact, I've mostly talked about how to make it less disagreeable and so implied that misery is the natural state of the learner. That has bothered me, because I'm actually much more in the "learning is fun" camp.

But perhaps a closer analysis shows that there's no inconsistency. Maybe learning is pleasurable when you pick the topic but a chore when someone else does. It made sense for me to write as though learning isn't enjoyable because I've focused on school-related tasks, which are assigned, not freely chosen. The strategies I've described would work just as well for content you choose to learn, but you probably didn't read this book for that purpose; you read it for help in learning the stuff you *have* to learn.

But how strong is the link between "have to learn" and "not fun"? Most students seem to think it's pretty consistent. Sure, sometimes you get lucky and an instructor assigns a book you actually enjoy, and sometimes a great instructor finds a way to intrigue you about a topic you initially didn't like. But even in those unusual cases, boredom or interest is still outside of your control.

The findings I've reviewed in this book indicate that that conclusion is mistaken. You can make yourself more interested in content that initially bores you. In this book, you've seen that:

- If information is interesting, you'll attend to it more closely.
- If you attend to it more closely, you will remember it better.
- If you remember it better, you will more likely do well on tests.
- If you do well on tests, you'll have more confidence in yourself as a student.
- If you're more confident, academic tasks will seem more achievable.
- If tasks seem more achievable, you'll procrastinate less.
- If you procrastinate less, you'll keep up with your work.

- If you keep up with your work, you'll know more about more topics.
- If you know something about a topic, new information on that subject will be easier to understand.
- If you understand new information, it will be more interesting.

My students know about the first three effects; they find it easy to study and remember stuff that interests them. They seldom consider the other effects and often don't know about some of them. For that reason, they view interest solely as a driver; they think that interest makes other processes (like attention and memory) work. They don't see that interest can be a product of other cognitive processes.

Here's the information from the list above displayed as a figure.

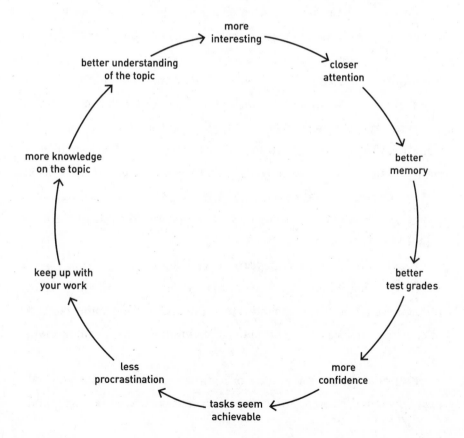

The figure makes it more obvious that you need not begin with interest. The components of learning form a virtuous cycle, and you can enter it anywhere, or at multiple points. In chapter 12 you learned ways of maintaining attention even when you're not very interested. In chapter 6 you learned how to improve your memory, in chapter 11 you learned how to overcome procrastination, and in chapter 13 I discussed different ways to think about self-confidence. As you change your self-confidence, your memory, your attention, and so on, the effects propagate around the circle, and your interest in what you're learning will increase.

You have probably experienced this effect firsthand. There was a subject you found boring and confusing, but you persisted until you understood it, and found that made it somewhat less boring. Perhaps even intriguing.

I think the definition of *independent learner* goes beyond the ability to acquire information and skills on your own when others demand it. It's also choosing what *you* want to learn. But how can you know what you want to learn if you don't know what's available to be learned?

Truly independent learners maintain a state of intellectual openness and curiosity. They are always ready to discover something new that they want to know more about. It's an optimistic way to live, because their curiosity is buttressed by the knowledge that new learning ultimately brings interest, enjoyment, and satisfaction. Anything that's unfamiliar can be a source of fun, and because each of us knows so little, the potential for fun is limitless.

People sometimes describe learning as "exploring new terrain" or as a "journey." I think the travel metaphor is apt; learning new things brings the same sense of adventure and satisfaction as traveling somewhere exotic, seeing the local flora and fauna, meeting the people, and observing how they live.

I set out to make the process of learning new information and skills easier, even in the absence of curiosity. To continue the travel metaphor, I hoped to create a map that ensured that you will arrive at the destination an instructor sets for you. But my fondest hope is that you'll think of this book as a travel kit and go exploring. I hope you'll be one of the people whose curiosity prompts them to see the world as brimming with hidden treasure.

ACKNOWLEDGMENTS

I've greatly benefited from ongoing conversations about teaching with the members of the Michigan Skills Project team: Colleen Counihan, Keith Desrosiers, Angela Duckworth, John Jonides, Ben Katz, Rhiannon Killian, Ethan Kross, and Ariana Orvell. I am also grateful to Miranda Beltzer, Katie Daniel, Jeremy Eberle, Nauder Namaky, Allie Silverman, Bethany Teachman, and especially Alex Werntz Czywczynski, who provided invaluable advice and feedback on the subject of anxiety.

I am deeply grateful to my editor, Karyn Marcus, and my agent, Esmond Harmsworth, for their substantial contributions in helping me to shape the presentation and also for their enthusiasm for this project. I owe a profound debt to two people with whom I've discussed most of the ideas presented here: David Daniel, to check my thinking on psychology, and Trisha Thompson-Willingham, to check my thinking on classroom realities and the probable reactions of students.

BIBLIOGRAPHY

CHAPTER 1

Bligh, Donald. *What's the Use of Lectures?* San Francisco: Jossey-Bass, 2000.

Cerbin, William. "Improving Student Learning from Lectures." *Scholarship of Teaching and Learning in Psychology* 4, no. 3 (June 2021): 151–63. http://dx.doi.org/10.1037/stl0000113.

deWinstanley, Patricia Ann, and Robert A. Bjork. "Successful Lecturing: Presenting Information in Ways That Engage Effective Processing." *New Directions for Teaching & Learning* 2002, no. 89 (Spring 2002): 19–31. https://doi.org/10.1002/tl.44.

Landrum, R. Eric. "Teacher-Ready Research Review: Clickers." *Scholarship of Teaching and Learning in Psychology* 1, no. 3 (September 2015): 250–54. https://doi.org/10.1037/stl0000031.

Plutarch. *Complete Works of Plutarch*. Hastings, East Sussex, UK: Delphi Classics, 2013 (e-book).

Raver, Sharon A., and Ann S. Maydosz. "Impact of the Provision and Timing of Instructor-Provided Notes on University Students' Learning." *Active Learning in Higher Education* 11, no. 3 (November 2010): 189–200. https://doi.org/10.1177/1469787410379682.

Shernoff, David J., Alexander J. Sannella, Roberta Y. Schorr, Lina Sanchez-Wall, Erik A. Ruzek, Suparna Sinha, and Denise M. Bressler.

"Separate Worlds: The Influence of Seating Location on Student Engagement, Classroom Experience, and Performance in the Large University Lecture Hall." *Journal of Environmental Psychology* 49 (April 2017): 55–64. https://doi.org/10.1016/J.JENVP.2016.12.002.

Worthington, Debra L., and David G. Levasseur. "To Provide or Not to Provide Course PowerPoint Slides? The Impact of Instructor-Provided Slides upon Student Attendance and Performance." *Computers & Education* 85 (July 2015): 14–22. https://doi.org/10.1016/j.compedu.2015.02.002.

CHAPTER 2

Carter, Susan Payne, Kyle Greenberg, and Michael S. Walker. "The Impact of Computer Usage on Academic Performance: Evidence from a Randomized Trial at the United States Military Academy." *Economics of Education Review* 56 (February 2017): 118–32. https://doi.org/10.1016/j.econedurev.2016.12.005.

Flanigan, Abraham E., and Scott Titsworth. "The Impact of Digital Distraction on Lecture Note Taking and Student Learning." *Instructional Science* 48, no. 5 (October 2020): 495–524. https://doi.org/10.1007/s11251-020-09517-2.

Gaudreau, Patrick, Dave Miranda, and Alexandre Gareau. "Canadian University Students in Wireless Classrooms: What Do They Do on Their Laptops and Does It Really Matter?" *Computers & Education* 70 (January 2014): 245–55. https://doi.org/10.1016/j.compedu.2013.08.019.

Luo, Linlin, Kenneth A. Kiewra, Abraham E. Flanigan, and Markeya S. Peteranetz. "Laptop Versus Longhand Note Taking: Effects on Lecture Notes and Achievement." *Instructional Science* 46, no. 6 (December 2018): 947–71. https://doi.org/10.1007/s11251-018-9458-0.

Mueller, Pam A., and Daniel M. Oppenheimer. "The Pen Is Mightier than the Keyboard: Advantages of Longhand over Laptop Note Taking." *Psychological Science* 25, no. 6 (June 2014): 1159–68. https://doi.org/10.1177/0956797614524581.

Peverly, Stephen T., Joanna K. Garner, and Pooja C. Vekaria. "Both Handwriting Speed and Selective Attention Are Important to Lecture Note-Taking." *Reading and Writing* 27, no. 1 (January 2014): 1–30. https://doi.org/10.1007/s11145-013-9431-x.

Peverly, Stephen T., Pooja C. Vekaria, Lindsay A. Reddington, James F. Sumowski, Kamauru R. Johnson, and Crystal M. Ramsay. "The Relationship of Handwriting Speed, Working Memory, Language Comprehension and Outlines to Lecture Note-Taking and Test-Taking Among College Students." *Applied Cognitive Psychology* 27, no. 1 (January–February 2013): 115–26. https://doi.org/10.1002/acp.2881.

Phillips, Natalie E., Brandon C. W. Ralph, Jonathan S. A. Carriere, and Daniel Smilek. "Examining the Influence of Saliency of Peer-Induced Distractions on Direction of Gaze and Lecture Recall." *Computers & Education* 99 (August 2016): 81–93. https://doi.org/10.1016/j.compedu.2016.04.006.

Piolat, Annie, Thierry Olive, and Ronald T. Kellogg. "Cognitive Effort During Note Taking." *Applied Cognitive Psychology* 19, no. 3 (April 2005): 291–312. https://doi.org/10.1002/acp.1086.

Reed, Deborah K., Hillary Rimel, and Abigail Hallett. "Note-Taking Interventions for College Students: A Synthesis and Meta-Analysis of the Literature." *Journal of Research on Educational Effectiveness* 9, no. 3 (January 2016): 307–33. https://doi.org/10.1080/19345747.2015.1105894.

Sana, Faria, Tina Weston, and Nicholas J. Cepeda. "Laptop Multitasking Hinders Classroom Learning for Both Users and Nearby Peers." *Computers & Education* 62 (March 2013): 24–31. https://doi.org/10.1016/j.compedu.2012.10.003.

Urry, Heather L., et al. "Don't Ditch the Laptop Just Yet: A Direct Replication of Mueller and Oppenheimer's (2014) Study 1 Plus Mini Meta-Analyses Across Similar Studies." *Psychological Science* 32, no. 3 (March 2021): 326–39. https://doi.org/10.1177/0956797620965541.

Will, Paris, Walter F. Bischof, and Alan Kingstone. "The Impact of Classroom Seating Location and Computer Use on Student Academic Performance." *PLOS ONE* 15, no. 8 (August 5, 2020): e0236131. https://doi.org/10.1371/journal.pone.0236131.

Williams, Andrew, Elisa Birch, and Phil Hancock. "The Impact of Online Lecture Recordings on Student Performance." *Australasian Journal of Educational Technology* 28, no. 2 (2012): 199–213. https://doi.org/10.14742/ajet.869.

CHAPTER 3

Brooks, Charles M., and Janice L. Ammons. "Free Riding in Group Projects and the Effects of Timing, Frequency, and Specificity of Criteria in Peer Assessments." *Journal of Education for Business* 78, no. 5 (May 2003): 268–72. https://doi.org/10.1080/08832320309598613.

Ericsson, Anders, and Robert Pool. *Peak: Secrets from the New Science of Expertise.* Boston: Houghton Mifflin Harcourt, 2016.

Ho, V. "Learning by Doing." In *Encyclopedia of Health Economics,* edited by Anthony J. Culyer, 141–45. Amsterdam: Elsevier, 2014.

Holyoak, Keith J., and Dušan Stamenković. "Metaphor Comprehension: A Critical Review of Theories and Evidence." *Psychological Bulletin* 144, no. 6 (June 2018): 641–71. https://doi.org/10.1037/bul0000145.

Long, Nicole, Brice A. Kuhl, and Marvin M. Chun. "Memory and Attention." In *Stevens' Handbook of Experimental Psychology and Cognitive Neuroscience,* 4th ed., Vol. 1, *Language and Memory,* edited by Elizabeth Phelps and Lila Davachi, 285–321. New York: Wiley, 2018.

Morris, C. Donald, John D. Bransford, and Jeffery J. Franks. "Levels of Processing Versus Transfer Appropriate Processing." *Journal of Verbal Learning and Verbal Behavior* 16, no. 5 (October 1977): 519–33. https://doi.org/10.1016/S0022-5371(77)80016-9.

Tulving, Endel, and Donald M. Thompson. "Encoding Specificity and Retrieval Processes in Episodic Memory." *Psychological Review* 80, no. 5 (September 1973): 352–73. https://doi.org/10.1037/h0020071.

CHAPTER 4

Bower, Gordon H., Michal C. Clark, Alan M. Lesgold, and David Winzenz. "Hiearchical Retrieval Schemes in Recall of Categorized Word Lists." *Journal of Verbal Learning and Verbal Behavior* 8, no. 3 (June 1969): 323–43. https://doi.org/10.1016/S0022-5371(69)80124-6.

Chularut, Pasana, and Teresa K. DeBacker. "The Influence of Concept Mapping on Achievement, Self-Regulation, and Self-Efficacy in Students of English as a Second Language." *Contemporary Educational Psychology* 29, no. 3 (July 2004): 248–63. https://doi.org/10.1016/j.cedpsych.2003.09.001.

Cohen, Dov, Emily Kim, Jacinth Tan, and Mary Ann Winkelmes. "A Note-Restructuring Intervention Increases Students' Exam Scores."

College Teaching 61, no. 3 (July 2013): 95–99. https://doi.org/10.1080/87567555.2013.793168.

Crawford, C. C. "The Correlation Between College Lecture Notes and Quiz Papers." *Journal of Educational Research* 12, no. 4 (1925): 282–91. https://doi.org/10.1080/00220671.1925.10879600.

Kiewra, Kenneth A. "Note Taking on Trial: A Legal Application of Note-Taking Research." *Educational Psychology Review* 28, no. 2 (June 2016): 377–84. https://doi.org/10.1007/s10648-015-9353-z.

Kiewra, Kenneth A., and Stephen L. Benton. "The Relationship Between Information-Processing Ability and Notetaking." *Contemporary Educational Psychology* 13, no. 1 (January 1988): 33–44. https://doi.org/10.1016/0361-476X(88)90004-5.

Luo, Linlin, Kenneth A. Kiewra, and Lydia Samuelson. "Revising Lecture Notes: How Revision, Pauses, and Partners Affect Note Taking and Achievement." *Instructional Science* 44 (February 2016): 45–67. https://doi.org/10.1007/s11251-016-9370-4.

Makany, Tamas, Jonathan Kemp, and Itiel E. Dror. "Optimising the Use of Note-Taking as an External Cognitive Aid for Increasing Learning." *British Journal of Educational Technology* 40, no. 4 (July 2009): 619–35. https://doi.org/10.1111/j.1467-8535.2008.00906.x.

Rachal, K. Chris, Sherri Daigle, and Windy S. Rachal. "Learning Problems Reported by College Students: Are They Using Learning Strategies?" *Journal of Instructional Psychology* 34, no. 4 (December 2007): 191–99. https://eric.ed.gov/?id=EJ790467.

CHAPTER 5

Bartoszewski, Brianna L., and Regan A. R. Gurung. "Comparing the Relationship of Learning Techniques and Exam Score." *Scholarship of Teaching and Learning in Psychology* 1, no. 3 (September 2015): 219–28. https://doi.org/10.1037/stl0000036.

Bohay, Mark, Daniel P. Blakely, Andrea K. Tamplin, and Gabriel A. Radvansky. "Note Taking, Review, Memory, and Comprehension." *American Journal of Psychology* 124, no. 1 (Spring 2011): 63–73. https://doi.org/10.5406/amerjpsyc.124.1.0063.

Denton, Carolyn A., Christopher A. Wolters, Mary J. York, Elizabeth Swanson, Paulina A. Kulesz, and David J. Francis. "Adolescents'

Use of Reading Comprehension Strategies: Differences Related to Reading Proficiency, Grade Level, and Gender." *Learning and Individual Differences* 37 (January 2015): 81–95. https://doi.org/10.1016 /j.lindif.2014.11.016.

Glenberg, Arthur M., Alex Cherry Wilkinson, and William Epstein. "The Illusion of Knowing: Failure in the Self-Assessment of Comprehension." *Memory & Cognition* 10, no. 6 (November 1982): 597–602. https://doi.org/10.3758/BF03202442.

Gurung, Regan A. R. "How Do Students Really Study (and Does It Matter)?" *Teaching of Psychology* 32, no. 4 (2005): 239–41.

Gurung, Regan A. R., and David B. Daniel. "Evidence-Based Pedagogy: Do Pedagogical Features Enhance Student Learning?" In *Best Practices for Teaching Introduction to Psychology*, edited by Dana S. Dunn and Stephen L. Chew, 41–55. Mahwah, NJ: Erlbaum, 2005.

Gurung, Regan A. R., Janet Weidert, and Amanda Jeske. "Focusing on How Students Study." *Journal of the Scholarship of Teaching and Learning* 10, no. 1 (January 2010): 28–35. https://files.eric.ed.gov/fulltext /EJ882123.pdf.

Jairam, Dharma, Kenneth A. Kiewra, Sarah Rogers-Kasson, Melissa Patterson-Hazley, and Kim Marxhausen. "SOAR Versus SQ3R: A Test of Two Study Systems." *Instructional Science* 42, no. 3 (May 2014): 409–20. https://doi.org/10.1007/s11251-013-9295-0.

McDaniel, Mark A., Daniel C. Howard, and Gilles O. Einstein. "The Read-Recite-Review Study Strategy: Effective and Portable." *Psychological Science* 20, no. 4 (April 2009): 516–22. https://doi.org/10.1111 /j.1467-9280.2009.02325.x.

Nist, Sherrie L., and Katie Kirby. "The Text Marking Patterns of College Students." *Reading Psychology* 10, no. 4 (Fall 1989): 321–38. https://doi .org/10.1080/0270271890100403.

Otero, José, and Walter Kintsch. "Failures to Detect Contradictions in a Text: What Readers Believe Versus What They Read." *Psychological Science* 3, no. 4 (July 1992): 229–35. https://doi.org/10.1111/j.1467 -9280.1992.tb00034.x.

Rayner, Keith, Elizabeth R. Schotter, Michael E. J. Masson, Mary C. Potter, and Rebecca Treiman. "So Much to Read, So Little Time: How Do We Read, and Can Speed Reading Help?" *Psychological*

Science in the Public Interest 17, no. 1 (May 2016): 4–34. https://doi .org/10.1177/1529100615623267.

CHAPTER 6

Alfieri, Louis, Timothy J. Nokes-Malach, and Christian D. Schunn. "Learning Through Case Comparisons: A Meta-analytic Review." *Educational Psychologist* 48, no. 2 (2013): 87–113. https://doi.org/10.10 80/00461520.2013.775712.

Baddeley, Alan D. *Your Memory: A User's Guide.* Buffalo, NY: Firefly, 2004.

Blasiman, Rachael N., John Dunlosky, and Katherine A. Rawson. "The What, How Much, and When of Study Strategies: Comparing Intended Versus Actual Study Behaviour." *Memory* 25, no. 6 (July 2017): 784–92. https://doi.org/10.1080/09658211.2016.1221974.

Callender, Aimee A., and Mark A. McDaniel. "The Limited Benefits of Rereading Educational Texts." *Contemporary Educational Psychology* 34, no. 1 (January 2009): 30–41. https://doi.org/10.1016/j .cedpsych.2008.07.001.

Catrambone, Richard. "The Subgoal Learning Model: Creating Better Examples to Improve Transfer to Novel Problems." *Journal of Experimental Psychology: General* 127, no. 4 (December 1998): 355–76. https:// doi.org/10.1037/0096-3445.127.4.355.

Dunlosky, John, Katherine A. Rawson, Elizabeth J. Marsh, Mitchell J. Nathan, and Daniel T. Willingham. "Improving Students' Learning with Effective Learning Techniques: Promising Directions from Cognitive and Educational Psychology." *Psychological Science in the Public Interest* 14, no. 1 (January 2013): 4–58. https://doi .org/10.1177/1529100612453266.

Fernandes, Myra A., Jeffrey D. Wammes, and Melissa E. Meade. "The Surprisingly Powerful Influence of Drawing on Memory." *Current Directions in Psychological Science* 27, no. 5 (October 2018): 302–308. https://doi.org/10.1177/0963721418755385.

Margulieux, Lauren E., and Richard Catrambone. "Improving Problem Solving with Subgoal Labels in Expository Text and Worked Examples." *Learning and Instruction* 42 (April 2016): 58–71. https://doi .org/10.1016/j.learninstruc.2015.12.002.

Rawson, Katherine A., John Dunlosky, and Sharon M. Sciartelli. "The Power of Successive Relearning: Improving Performance on Course Exams and Long-Term Retention." *Educational Psychology Review* 25, no. 4 (December 2013): 523–48. https://doi.org/10.1007/s10648-013 -9240-4.

Tauber, Sarah K., Amber E. Witherby, John Dunlosky, Katherine A. Rawson, Adam L. Putnam, and Henry L. Roediger III. "Does Covert Retrieval Benefit Learning of Key-Term Definitions?" *Journal of Applied Research in Memory and Cognition* 7, no. 1 (March 2018): 106–15. https://doi.org/10.1016/j.jarmac.2016.10.004.

Willingham, Daniel T. "Does Tailoring Instruction to 'Learning Styles' Help Students Learn?" *American Educator* 42, no. 2 (Summer 2018): 28–36. https://files.eric.ed.gov/fulltext/EJ1182080.pdf.

Yang, Chunliang, Liang Luo, Miguel A. Vadillo, Rongjun Yu, and David R. Shanks. "Testing (Quizzing) Boosts Classroom Learning: A Systematic and Meta-analytic Review." *Psychological Bulletin* 147, no. 4 (April 2021): 399–435. https://doi.org/10.1037/bul0000309.

CHAPTER 7

Bjork, Elizabeth L., and Robert A. Bjork. "Making Things Hard on Yourself, but in a Good Way: Creating Desirable Difficulties to Enhance Learning." In *Psychology and the Real World: Essays Illustrating Fundamental Contributions to Society*, edited by Morton A. Gernsbacher, Richard W. Pew, Leaetta M. Hough, and James R. Pomerantz, 56–64. New York: FABBS Foundation, 2009.

Dougherty, Kathleen M., and James M. Johnston. "Overlearning, Fluency, and Automaticity." *Behavior Analyst* 19, no. 2 (October 1996): 289–92. https://doi.org/10.1007/BF03393171.

Hertzog, Christopher, Jarrod C. Hines, and Dayna R. Touron. "Judgments of Learning Are Influenced by Multiple Cues in Addition to Memory for Past Test Accuracy." *Archives of Scientific Psychology* 1, no. 1 (2013): 23–32. https://doi.org/10.1037/arc0000003.

Kornell, Nate, and Hannah Hausman. "Performance Bias: Why Judgments of Learning Are Not Affected by Learning." *Memory & Cognition* 45, no. 8 (November 2017): 1270–80. https://doi.org/10.3758 /s13421-017-0740-1.

Roelle, Julian, Elisabeth Marie Schmidt, Alica Buchau, and Kirsten Berthold. "Effects of Informing Learners About the Dangers of Making Overconfident Judgments of Learning." *Journal of Educational Psychology* 109, no. 1 (January 2017): 99–117. https://doi.org/10.1037/edu0000132.

Schwartz, Bennett L., and Janet Metcalfe. "Metamemory: An Update of Critical Findings." In *Cognitive Psychology of Memory*, Vol. 2, *A Comprehensive Reference*, edited by J. H. Byrne, 423–32. Oxford, UK: Academic Press, 2017.

Shanks, Lindzi L., and Michael J. Serra. "Domain Familiarity as a Cue for Judgments of Learning." *Psychonomic Bulletin & Review* 21, no. 2 (April 2014): 445–53. https://doi.org/10.3758/s13423-013-0513-1.

Soderstrom, Nicholas C., and Robert A. Bjork. "Learning Versus Performance: An Integrative Review." *Perspectives on Psychological Science* 10, no. 2 (March 2015): 176–99. https://doi.org/10.1177/1745691615569000.

Witherby, Amber E., and Sarah K. Tauber. "The Influence of Judgments of Learning on Long-Term Learning and Short-Term Performance." *Journal of Applied Research in Memory and Cognition* 6, no. 4 (December 2017): 496–503. https://doi.org/10.1016/j.jarmac.2017.08.004.

CHAPTER 8

Archer, N. Sidney, and Ralph Pippert. "Don't Change the Answer!: An Exposé of the Perennial Myth That the First Choices Are Always the Correct Ones." *The Clearing House: A Journal of Educational Strategies, Issues and Ideas* 37, no. 1 (1962): 39–41. https://doi.org/10.1080/00098655.1962.11476207.

Bourassa, Kyle J., John M. Ruiz, and David A. Sbarra. "The Impact of Physical Proximity and Attachment Working Models on Cardiovascular Reactivity: Comparing Mental Activation and Romantic Partner Presence." *Psychophysiology* 56, no. 5 (May 2019): 1–12. https://doi.org/10.1111/psyp.13324.

Bramão, Inês, Anna Karlsson, and Mikael Johansson. "Mental Reinstatement of Encoding Context Improves Episodic Remembering." *Cortex* 94 (September 2017): 15–26. https://doi.org/10.1016/j.cortex.2017.06.007.

Calma-Birling, Destany, and Regan A. R. Gurung. "Does a Brief Mindfulness Intervention Impact Quiz Performance?" *Psychology*

Learning & Teaching 16, no. 3 (November 2017): 323–35. https://doi
.org/10.1177/1475725717712785.

Copeland, David A. "Should Chemistry Students Change Answers on
Multiple-Choice Tests?" *Journal of Chemical Education* 49, no. 4 (April
1972): 258. https://doi.org/10.1021/ed049p258.

DiBattista, David, Jo-Anne Sinnige-Egger, and Glenda Fortuna. "The
'None of the Above' Option in Multiple-Choice Testing: An Experimental Study." *Journal of Experimental Education* 82, no. 2 (2014): 168–
83. https://doi.org/10.1080/00220973.2013.795127.

Embse, Nathaniel von der, Justin Barterian, and Natasha Segool. "Test
Anxiety Interventions for Children and Adolescents: A Systematic Review of Treatment Studies from 2000–2010." *Psychology in
the Schools* 50, no. 1 (January 2013): 57–71. https://doi.org/10.1002
/PITS.21660.

Erdelyi, Matthew Hugh. *The Recovery of Unconscious Memories: Hyper-
mnesia and Reminiscence.* Chicago: University of Chicago Press, 1996.

Kruger, Justin, Derrick Wirtz, and Dale T. Miller. "Counterfac-
tual Thinking and the First Instinct Fallacy." *Journal of Personality and Social Psychology* 88, no. 5 (May 2005): 725–35. https://doi
.org/10.1037/0022-3514.88.5.725.

Pichert, James W., and Richard C. Anderson. "Taking Different Perspectives on a Story." *Journal of Educational Psychology* 69, no. 4 (August
1977): 309–315. https://doi.org/10.1037/0022-0663.69.4.309.

Schwarz, Shirley P., Robert F. McMorris, and Lawrence P. DeMers.
"Reasons for Changing Answers: An Evaluation Using Personal In-
terviews." *Journal of Educational Measurement* 28, no. 2 (June 1991):
163–71. https://doi.org/10.1111/j.1745-3984.1991.tb00351.x.

Smith, Steven M., and Justin D. Handy. "Effects of Varied and Constant
Environmental Contexts on Acquisition and Retention." *Journal of
Experimental Psychology: Learning, Memory, and Cognition* 40, no. 6
(November 2014): 1582–93. https://doi.org/10.1037/xlm0000019.

Vispoel, Walter P. "Reviewing and Changing Answers on Com-
puterized Fixed-Item Vocabulary Tests." *Educational and Psychological Measurement* 60, no. 3 (June 2000): 371–84. https://doi.org
/10.1177/00131640021970600.

CHAPTER 9

Black, Paul, and Dylan Wiliam. "Developing the Theory of Formative Assessment." *Educational Assessment, Evaluation and Accountability* 21, no. 5 (2009): 5–31. https://doi.org/10.1007/s11092-008-9068-5.

Dweck, Carol S. *Mindset: Changing the Way You Think to Fulfil Your Potential*, 6th ed. New York: Random House, 2017.

Kornell, Nate, and Janet Metcalfe. "The Effects of Memory Retrieval, Errors and Feedback on Learning." In *Applying Science of Learning in Education: Infusing Psychological Science into the Curriculum*, edited by Victor A. Benassi, Catherine E. Overson, and Christopher M. Hakala, 225–51. Washington, DC: Division 2, American Psychological Association, 2014. http://www.columbia.edu/cu/psychology/metcalfe/PDFs/Kornell2013.pdf.

Mayer, Richard E. "Rote Versus Meaningful Learning." *Theory into Practice* 41, no. 4 (Autumn 2002): 226–32. https://doi.org/10.1207/s15430421tip4104_4.

Morrison, Frederick J., Matthew H. Kim, Carol M. Connor, and Jennie K. Grammer. "The Causal Impact of Schooling on Children's Development: Lessons for Developmental Science." *Current Directions in Psychological Science* 28, no. 5 (October 2019): 441–49. https://doi.org/10.1177/0963721419855661.

Roberts, Dennis M. "An Empirical Study on the Nature of Trick Test Questions." *Journal of Educational Measurement* 30, no. 4 (December 1993): 331–44. https://doi.org/10.1111/J.1745-3984.1993.TB00430.X.

Shute, Valerie J. "Focus on Formative Feedback." *Review of Educational Research* 78, no. 1 (March 2008): 153–89. https://doi.org/10.3102/0034654307313795.

Simons, Daniel J., Walter R. Boot, Neil C. Charness, Susan E. Gathercole, Christopher F. Chabris, David Z. Hambrick, and Elizabeth A. L. Stine-Morrow. "Do 'Brain Training' Programs Work?" *Psychological Science in the Public Interest* 17, no. 3 (October 2016): 103–86.

CHAPTER 10

Astill, Rebecca G., Kristiaan B. Van der Heijden, Marinus H. Van IJzendoorn, and Eus J. W. Van Someren. "Sleep, Cognition, and Behavioral Problems in School-Age Children: A Century of Research Meta-

Analyzed." *Psychological Bulletin* 138, no. 6 (November 2012): 1109–38. https://doi.org/10.1037/a0028204.

Buehler, Roger, Dale Griffin, and Michael Ross. "Exploring the 'Planning Fallacy': Why People Underestimate Their Task Completion Times." *Journal of Personality and Social Psychology* 67, no. 3 (September 1994): 366–81. https://doi.org/10.1037/0022-3514.67.3.366.

Cain, Neralie, and Michael Gradisar. "Electronic Media Use and Sleep in School-Aged Children and Adolescents: A Review." *Sleep Medicine* 11, no. 8 (September 2010): 735–42. https://doi.org/10.1016/j.sleep.2010.02.006.

Crovitz, Herbert F., and Walter F. Daniel. "Measurements of Everyday Memory: Toward the Prevention of Forgetting." *Bulletin of the Psychonomic Society* 22, no. 5 (November 1984): 413–14. https://doi.org/10.3758/BF03333861.

Gollwitzer, Peter M., Fujita Kentaro, and Gabriele Oettingen. "Planning and the Implementation of Goals." In *Handbook of Self-Regulation: Research, Theory, and Applications*, edited by Roy F. Baumeister and Kathleen D. Vohs, 211–28. New York: Guilford, 2004.

Gomez Fonseca, Angela, and Lisa Genzel. "Sleep and Academic Performance: Considering Amount, Quality and Timing." *Current Opinion in Behavioral Sciences* 33 (June 2020): 65–71. https://doi.org/10.1016/j.cobeha.2019.12.008.

Hartwig, Marissa K., and John Dunlosky. "Study Strategies of College Students: Are Self-Testing and Scheduling Related to Achievement?" *Psychonomic Bulletin & Review* 19, no. 1 (February 2012): 126–34. https://doi.org/10.3758/s13423-011-0181-y.

Kornell, Nate, and Robert A. Bjork. "The Promise and Perils of Self-Regulated Study." *Psychonomic Bulletin & Review* 14, no. 2 (April 2007): 219–24. https://doi.org/10.3758/bf03194055.

Krause, Adam J., Eti Ben Simon, Bryce A. Mander, Stephanie M. Greer, Jared M. Saletin, Andrea N. Goldstein-Piekarski, and Matthew P. Walker. "The Sleep-Deprived Human Brain." *Nature Reviews Neuroscience* 18, no. 7 (July 2017): 404–18. https://doi.org/10.1038/nrn.2017.55.

Kross, Ethan, Emma Bruehlman-Senecal, Jiyoung Park, Aleah Burson, Adrienne Dougherty, Holly Shablack, Ryan Bremner, Jason Moser, and Ozlem Ayduk. "Self-Talk as a Regulatory Mechanism: How You

Do It Matters." *Journal of Personality and Social Psychology* 106, no. 2 (February 2014): 304–24. https://doi.org/10.1037/a0035173.

Shirtcliff, Elizabeth A., Amber L. Allison, Jeffrey M. Armstrong, Marcia J. Slattery, Ned H. Kalin, and Marilyn J. Essex. "Longitudinal Stability and Developmental Properties of Salivary Cortisol Levels and Circadian Rhythms from Childhood to Adolescence." *Developmental Psychobiology* 54, no. 5 (July 2012): 493–502. https://doi.org/10.1002/dev.20607.

CHAPTER 11

Ariely, Dan, and Klaus Wertenbroch. "Procrastination, Deadlines, and Performance: Self-Control by Precommitment." *Psychological Science* 13, no. 3 (May 2002): 219–24. https://doi.org/10.1111/1467-9280.00441.

Barrera, Manuel, Jr., and Sheila L. Ainlay. "The Structure of Social Support: A Conceptual and Empirical Analysis." *Journal of Community Psychology* 11, no. 2 (April 1983): 133–43. https://doi.org/10.1002/1520-6629(198304)11:2<133::AID-JCOP2290110207>3.0.CO;2-L.

Frederick, Shane, Nathan Novemsky, Jing Wang, Ravi Dhar, and Stephen Nowlis. "Opportunity Cost Neglect." *Journal of Consumer Research* 36, no. 4 (December 2009): 553–61. https://doi.org/10.1086/599764.

Krause, Kathrin, and Alexandra M. Freund. "It's in the Means: Process Focus Helps Against Procrastination in the Academic Context." *Motivation and Emotion* 40, no. 3 (June 2016): 422–37. https://doi.org/10.1007/s11031-016-9541-2.

Lally, Phillippa, Cornelia H. M. van Jaarsveld, Henry W. W. Potts, and Jane Wardle. "How Are Habits Formed: Modelling Habit Formation in the Real World." *European Journal of Social Psychology* 40, no. 6 (October 2010): 998–1009. https://doi.org/10.1002/EJSP.674.

Lickel, Brian, Kostadin Kushlev, Victoria Savalei, Shashi Matta, and Toni Schmader. "Shame and the Motivation to Change the Self." *Emotion* 14, no. 6 (December 2014): 1049–61. https://doi.org/10.1037/A0038235.

Neal, David T., Wendy Wood, Jennifer S. Labrecque, and Phillippa Lally. "How Do Habits Guide Behavior? Perceived and Actual Triggers of Habits in Daily Life." *Journal of Experimental Social Psychology* 48, no. 2 (March 2012): 492–98. https://doi.org/10.1016/j.jesp.2011.10.011.

Ruby, Matthew B., Elizabeth W. Dunn, Andrea Perrino, Randall Gillis, and Sasha Viel. "The Invisible Benefits of Exercise." *Health Psychology* 30, no. 1 (January 2011): 67–74. https://doi.org/10.1037/a0021859.

Steel, Piers. "The Nature of Procrastination: A Meta-Analytic and Theoretical Review of Quintessential Self-Regulatory Failure." *Psychological Bulletin* 133, no. 1 (January 2007): 65–94. https://doi.org/10.1037/0033-2909.133.1.65.

Strunk, Kamden K., and Misty R. Steele. "Relative Contributions of Self-Efficacy, Self-Regulation, and Self-Handicapping in Predicting Student Procrastination." *Psychological Reports* 109, no. 3 (December 2011): 983–89. https://doi.org/10.2466/07.09.20.PR0.109.6.983-989.

Wood, Wendy, and Dennis Rünger. "Psychology of Habit." *Annual Review of Psychology* 67 (January 2016): 289–314. https://doi.org/10.1146/annurev-psych-122414-033417.

Zhang, Shunmin, Peiwei Liu, and Tingyong Feng. "To Do It Now or Later: The Cognitive Mechanisms and Neural Substrates Underlying Procrastination." *WIREs Cognitive Science* 10, no. 4 (July–August 2019). https://doi.org/10.1002/WCS.1492.

CHAPTER 12

Allen, Andrew P., and Andrew P. Smith. "A Review of the Evidence That Chewing Gum Affects Stress, Alertness and Cognition." *Journal of Behavioral and Neuroscience Research* 9, no. 1 (2011): 7–23. https://www.researchgate.net/publication/313065290_A_review_of_the_evidence_that_chewing_gum_affects_stress_alertness_and_cognition.

Altmann, Erik M., J. Gregory Trafton, and David Z. Hambrick. "Momentary Interruptions Can Derail the Train of Thought." *Journal of Experimental Psychology: General* 143, no. 1 (February 2014): 215–26. https://doi.org/10.1037/a0030986.

Ariga, Atsunori, and Alejandro Lleras. "Brief and Rare Mental 'Breaks' Keep You Focused: Deactivation and Reactivation of Task Goals Preempt Vigilance Decrements." *Cognition* 118, no. 3 (March 2011): 439–43. https://doi.org/10.1016/j.cognition.2010.12.007.

Berridge, Kent C. "Wanting and Liking: Observations from the Neuroscience and Psychology Laboratory." *Inquiry* 52, no. 4 (August 2009): 378–98. https://doi.org/10.1080/00201740903087359.

Bratman, Gregory N., J. Paul Hamilton, and Gretchen C. Daily. "The Impacts of Nature Experience on Human Cognitive Function and Mental Health." *Annals of the New York Academy of Sciences* 1249, no. 1 (February 2012): 118–36. https://doi.org/10.1111/j.1749-6632.2011.06400.x.

Duckworth, Angela L., and James J. Gross. "Behavior Change." *Organizational Behavior and Human Decision Processes* 161 (suppl.) (November 2020): 39–49. https://doi.org/10.1016/j.obhdp.2020.09.002.

Duckworth, Angela L., Tamar Szabó Gendler, and James J. Gross. "Situational Strategies for Self-Control." *Perspectives on Psychological Science* 11, no. 1 (January 2016): 35–55. https://doi.org/10.1177/1745691615623247.

Junco, Reynol, and Shelia R. Cotten. "The Relationship Between Multitasking and Academic Performance." *Computers & Education* 59, no. 2 (September 2012): 505–14. https://doi.org/10.1016/j.compedu.2011.12.023.

Onyper, Serge V., Timothy L. Carr, John S. Farrar, and Brittney R. Floyd. "Cognitive Advantages of Chewing Gum. Now You See Them, Now You Don't." *Appetite* 57, no. 2 (October 2011): 321–28. https://doi.org/10.1016/j.appet.2011.05.313.

Orvell, Ariana, Brian D. Vickers, Brittany Drake, Philippe Verduyn, Ozlem Ayduk, Jason Moser, John Jonides, and Ethan Kross. "Does Distanced Self-Talk Facilitate Emotion Regulation Across a Range of Emotionally Intense Experiences?" *Clinical Psychological Science* 9, no. 1 (January 2021): 68–78. https://doi.org/10.1177/2167702620951539.

Rideout, Victoria, and Michael B. Robb. "The Common Sense Census: Media Use by Tweens and Teens." San Francisco: Common Sense Media, 2019. https://www.commonsensemedia.org/sites/default/files/uploads/research/2019-census-8-to-18-full-report-updated.pdf.

Risko, Evan F., Dawn Buchanan, Srdan Medimorec, and Alan Kingstone. "Everyday Attention: Mind Wandering and Computer Use during Lectures." *Computers & Education* 68 (October 2013): 275–83. https://doi.org/10.1016/j.compedu.2013.05.001.

Seli, Paul, Evan F. Risko, Daniel Smilek, and Daniel L. Schacter. "Mind-Wandering with and Without Intention." *Trends in Cognitive Sciences* 20, no. 3 (August 2016): 605–17. https://doi.org/10.1016/j.tics.2016.05.010.

Smallwood, Jonathan, and Jonathan W. Schooler. "The Science of Mind Wandering: Empirically Navigating the Stream of Consciousness." *Annual Review of Psychology* 66 (January 2015): 487–518. https://doi .org/10.1146/annurev-psych-010814-015331.

Stothart, Cary, Ainsley Mitchum, and Courtney Yehnert. "The Attentional Cost of Receiving a Cell Phone Notification." *Journal of Experimental Psychology: Human Perception and Performance* 41, no. 4 (August 2015): 893–97. https://doi.org/10.1037/xhp0000100.

Uncapher, Melina R., and Anthony D. Wagner. "Minds and Brains of Media Multitaskers: Current Findings and Future Directions." *Proceedings of the National Academy of Sciences of the United States of America* 115, no. 40 (October 2018): 9889–96. https://doi.org/10.1073 /pnas.1611612115.

Willingham, Daniel T. "The High Price of Multitasking." *New York Times*, July 14, 2019. https://www.nytimes.com/2019/07/14/opinion /multitasking-brain.html.

Zanesco, Anthony P., Brandon G. King, Katherine A. MacLean, Tonya L. Jacobs, Stephen R. Aichele, B. Alan Wallace, Jonathan Smallwood, Jonathan W. Schooler, and Clifford D. Saron. "Meditation Training Influences Mind Wandering and Mindless Reading." *Psychology of Consciousness: Theory, Research and Practice* 3, no. 1 (March 2016): 12–33. https://doi.org/10.1037/cns0000082.

CHAPTER 13

Arens, Katrin A., Herbert W. Marsh, Reinhard Pekrun, Stephanie Lichtenfeld, Kou Murayama, and Rudolf vom Hofe. "Math Self-Concept, Grades, and Achievement Test Scores: Long-Term Reciprocal Effects Across Five Waves and Three Achievement Tracks." *Journal of Educational Psychology* 109, no. 5 (July 2017): 621–34. https://doi .org/10.1037/edu0000163.

Brummelman, Eddie, and Sander Thomaes. "How Children Construct Views of Themselves: A Social-Developmental Perspective." *Child Development* 88, no. 6 (November–December 2017): 1763–73. https:// doi.org/10.1111/cdev.12961.

Dedonno, Michael A. "The Influence of Family Attributes on College Students' Academic Self-Concept." *North American Journal of Psy-*

chology 15, no. 1 (March 2013): 49–62. https://www.researchgate.net
/publication/235986598.

Koivuhovi, Satu, Herbert W. Marsh, Theresa Dicke, Baljinder Sahdra,
Jiesi Guo, Philip D. Parker, and Mari-Pauliina Vainikainen. "Academic
Self-Concept Formation and Peer-Group Contagion: Development of
the Big-Fish-Little-Pond Effect in Primary-School Classrooms and
Peer Groups." *Journal of Educational Psychology* (advance online publi-
cation). https://doi.org/10.1037/edu0000554.

Marsh, Herbert W. "Academic Self-Concept: Theory, Measurement,
and Research." In *Psychological Perspectives on the Self*, Vol. 4, *The Self
in Social Perspective*, edited by Jerry Suls, 71–110. East Sussex, UK:
Psychology Press, 2016.

Marsh, Herbert W., Reinhard Pekrun, Kou Murayama, A. Katrin Ehrens,
Philip D. Parker, Jiesi Guo, and Theresa Dicke. "An Integrated Model
of Academic Self-Concept Development: Academic Self-Concept,
Grades, Test Scores, and Tracking over 6 Years." *Developmental Psychology* 54, no. 2 (February 2018): 263–80. https://doi.org/10.1037
/dev0000393.

Tan, Cheng Yong, Meiyan Lyu, and Baiwen Peng. "Academic Benefits
from Parental Involvement Are Stratified by Parental Socioeconomic
Status: A Meta-analysis." *Parenting: Science and Practice* 20, no. 4
(2020): 241–87. https://doi.org/10.1080/15295192.2019.1694836.

Wolff, Fabian, Friederike Helm, Friederike Zimmermann, Gabriel Nagy,
and Jens Möller. "On the Effects of Social, Temporal, and Dimensional
Comparisons on Academic Self-Concept." *Journal of Educational Psychology* 110, no. 7 (October 2018): 1005–25. https://doi.org/10.1037
/EDU0000248.

CHAPTER 14

Bernstein, Douglas A., Bethany A. Teachman, Bunmi O. Olatunji, and
Scott O. Lilienfeld. "Cognitive, Behavioral, and Acceptance-Based
Psychotherapies." In *Introduction to Clinical Psychology: Bridging Science
and Practice*, 286–323. Cambridge, UK: Cambridge University Press,
2020.

Credé, Marcus, and Nathan R. Kuncel. "Study Habits, Skills, and Attitudes: The Third Pillar Supporting Collegiate Academic Performance."

Perspectives on Psychological Science 3, no. 6 (November 2008): 425–53. https://doi.org/10.1111/j.1745-6924.2008.00089.x.

Duits, Puck, Danielle C. Cath, Shmuel Lissek, Joop J. Hox, Alfons O. Hamm, Iris M. Engelhard, Marcel A. van den Hout, and Joke M. P. Baas. "Updated Meta-analysis of Classical Fear Conditioning in the Anxiety Disorders." *Depression & Anxiety* 32, no. 4 (April 2015): 239–53. https://doi.org/10.1002/DA.22353.

Eysenck, Michael W. *Anxiety: The Cognitive Perspective.* Hove, UK: Lawrence Erlbaum Associates, 1992.

Hirsch, Colette R., and Andrew Mathews. "A Cognitive Model of Pathological Worry." *Behaviour Research and Therapy* 50, no. 10 (October 2012): 636–46. https://doi.org/10.1016/j.brat.2012.06.007.

Hirsch, Colette R., Charlotte Krahé, Jessica Whyte, Sofia Loizou, Livia Bridge, Sam Norton, and Andrew Mathews. "Interpretation Training to Target Repetitive Negative Thinking in Generalized Anxiety Disorder and Depression." *Journal of Consulting and Clinical Psychology* 86, no. 12 (December 2018): 1017–30. https://doi.org/10.1037/CCP0000310.

Hoge, Elizabeth A., Eric Bui, Luana Marques, Christina A. Metcalf, Laura K. Morris, Donald J. Robinaugh, John J. Worthington, Mark H. Pollack, and Naomi M. Simon. "Randomized Controlled Trial of Mindfulness Meditation for Generalized Anxiety Disorder: Effects on Anxiety and Stress Reactivity." *Journal of Clinical Psychiatry* 74, no. 8 (August 2013): 786–92. https://doi.org/10.4088/JCP.12M08083.

Stein, Murray B., Kerry L. Jang, and W. John Livesley. "Heritability of Anxiety Sensitivity: A Twin Study." *Journal of Psychiatry* 156, no. 2 (February 1999): 246–51. https://ajp.psychiatryonline.org/doi/pdf/10.1176/ajp.156.2.246.

INDEX

Note: Page references in italics indicate illustrations, and t *indicates a table.*

activities, learning from, 50–73
  analogies, 56–58
  attention to features of activities,
    52–53
  concentration, 65
  for experience, 51, 66–69
  experience vs. practice, 63–66
  feedback, 62, 65, 70–71
  for instructors, 71–73
  the instructor's perspective, 69–71
  mapping, 57–58
  note taking, 55, 68–69
  participation, 55
  preparation, 55, 71
  presence and engagement, 55–56
  for process learning, 51, 62
  project selection/planning, 61–62,
    72
  purposes of, 51–52
  reflection on what you learned,
    62–63
  science labs, 59–60, 71–72
  scripted activities, 59–61, 71
  for skill improvement, 64–66
  for understanding, 51
Adams, Scott, 263–64
analogies, 56–58
Anderson, Richard, 157–58
anxiety, 273–87
  accommodations for a student
    with, 286–87
  avoiding common responses to,
    279–80
  benefits of, 273
  catastrophizing, 279
  as damaging, 273–74
  denying/suppressing, 280
  about exams, 152–54, 273, 287
  as excitement, 283
  genetic predisposition for, 274
  giving up, 279
  for instructors, 286–87
  irrational, 274–75
  learned, 274

anxiety *(cont.)*,
    medication for, 280
    meditation for, 284–86
    prevalence of, 286
    reinterpreting anxious thoughts,
        281–82
    reinterpreting bodily signs, 283
    success as doing a little of what
        you want to do, 277–78
    unlearning old associations,
        275–76
Aristotle, 51, 64
attention, 52–54, 140
Augustine, Saint, 137

caffeine, 153
calendar use, 201–3
Callender, Aimee, 109
Catrambone, Richard, 130
Centers for Disease Control and
    Prevention, 196
Churchill, Winston, 207
concentration training programs,
    253
cortisol, 196
curse of knowledge, 53–54

Daniel, David, 102, 168
distractions. *See* focus
Duckworth, Angela, 237

80/20 rule, 168
enjoyment, research on, 248–49
exam results, evaluating, 174–92

analyzing questions you got right,
    186–87
categorizing your mistakes,
    175–78
essay questions, analyzing
    mistakes on, 179–82
failed exams and intelligence,
    187–89
feedback from the instructor, 179,
    190–91
for instructors, 190–92
reluctance to analyze mistakes,
    187, 189–90
trick questions, 183–85
exams, readiness for, 135–48
ability to explain content, 138, 147
for instructors, 146–48
"knowing" something, meaning of,
    137–39
multiple-choice vs. short-answer
    vs. essay tests, 148
overconfidence about your
    knowledge, 136–37, 139–41
overlearning as protection against
    forgetting, 145–46
performance vs. learning, 136,
    138–39, 145
practice tests, 143–44
rereading, 140–41
self-testing, 141–43
exams, studying for, 105–34
application problems, 128–33
cramming, 125–27, 198–99
found materials, 115–16, 133
for instructors, 133–34
labeling subgoals of problem
    variations, 130–33

and learning style, 122–23

memorization strategies for, 107–9

mnemonics, 118–20

posing/answering questions for memory retention, 112–14, 116–18

retrieval practice, 106, 108, 110–11

study groups, 112, 124

study guides (flash cards), 111–14, 120–22

study preparation as studying, 109–11

work problems, 130–31

exams, taking, 149–73

anxiety about, 152–54, 273, 287

asking the instructor for clarification, 164–66

changing an answer, 161

checking your work, 151–52

dubious strategies for, 149–50, 166–68

essay questions, 168–71

fill-in-the-blank questions, 171

for instructors, 171–73

keep trying to remember, 160–62

multiple-choice questions, 149–50, 152, 162–63, 166–68, 171

overthinking, 166–68

pop knowledge, 162–64

reading instructions and questions carefully, 151

short-answer questions, 171

skimming the test and calculating time needed, 151

themes' use in searching memory, 157–59

visualizing your study environment, 155–57

exercise, 226, 253

fight-or-flight response, 283

flash cards. See study guides

focus, 237–58

breaks, 254–56

distractions, avoiding, 240, 244–46

distractions, reevaluating, 246–48, 256–57

distractions, stages of, 237–38, 246, 258

gum chewing, 250–51

for instructors, 257–58

meditation, 252–54

mind wandering, 238, 247, 251–54

moving along, 256–57

while multitasking, 244–46, 257–58

by reading aloud, 251–52

regrouping (evaluating your task and methods), 256

and social media, enjoying vs. wanting, 248–50

to-do lists, 252, 255

work environment, choosing, 239–41

work environment, improving, 241–43

FOMO (fear of missing out), 232

Freund, Alexandra, 225

goals
  ambitious vs. small, 221
  and emotions, 211
  and environment, 210–11
  factors in setting, 209–12
  following your passion, 209–10
  long-term, 206–9, 214–15
  making specific plans to meet,
    212–13
Gross, James, 237
Gurung, Regan, 102

habits, 218–20

independent learning, 1–2,
  292
intelligence, 187–89, 231
isolating during exams, 153

knowledge
  as a curse, 53–54
  overconfidence about, 136–37,
    139–41
  pop, 162–64, 184–86

Lamott, Anne, 222
learning
  attention needed for, 52–53
  desire to learn, 4
  by doing (see activities, learning
    from)
  vs. exercise, 5
  as fun, 290

"good learner," meaning of,
  262–64
independent, 1–2, 292
as interesting, 290–92, 291
as a journey, 292–93
vs. performance, 136, 138–39, 145
remote, 89
by repetition, 4
by speaking out loud, 120–21
speed of, 188, 262
styles of, 122–23
by teaching, 121
lectures
  active thinking and listening,
    15–17, 24–25
  activities interspersed into (see
    activities, learning from)
  asking questions, 15, 21–25
  the big picture, 14–15
  vs. conversation, 9–10, 16
  feedback to the instructor, 23
  for instructors, 24–26
  mental processes required to
    attend, 28
  notes from the instructor, 17–19
  organization of, 9–10, 11, 12–15,
    24
  reading assignments, 19–20
  recognizing when you don't
    understand something, 7–8, 25
  recorded vs. live, 46–47, 89
  See also note taking during lectures
listening, 15–17, 24–25

McDaniel, Mark, 109
meditation, 153, 252–54, 284–86

memory
  and attention, 53–54, 140
  environment's connection with,
    155–57
  familiarity process, 140
  improving via probing, 105
  mental walks, 119
  mnemonics, 118–20
  organization as aiding, 75, 75–77,
    105, 108
  prospective, 193–94
  as a residue of thought, 52–53, 59,
    105, 108, 129, 156
  retrieval practice, 106, 108,
    110–11, 160
  sleep's effect on, 196
  and studying for exams (*see under*
    exams, studying for)
  themes' use in searching, 157–59
mindfulness-based stress reduction
  (MBSR), 285
mindfulness meditation, 253–54,
  284–86
mind wandering, 238, 247, 251–
  54
multitasking, 244–46, 257–58
music while working, 245–46

napping, 198
notes, reorganizing, 74–89
  connections among elements of
    notes, 77–79
  copying/beautifying your notes,
    85–86
  to identify missing information,
    80–81, 86–88, 87–88*t*

and the impulse toward novelty,
    76
  for instructors, 86–89
  instructor's help with, 84–85
  memory aided by, 75, 75–77
  in a study group, 82–83, 87
  as studying, 77, 83
  tree diagrams, 78, 80
  to understand how the lecture
    relates to the readings, 79
note taking during activities, 55,
  68–69
note taking during lectures, 27–49
  abbreviations and symbols, 41–43,
    45
  on alternate pages, 41
  color coding of, 31
  coordinating your notes with the
    instructor's, 18–19
  evaluating your notes on the spot,
    38–39
  figures and graphs, 44, 44–45
  for instructors, 47–49
  on a laptop, 31, 35–37, 48–49
  and lecture recordings, 46–47
  in longhand, 35–37, 48
  mental processes needed for,
    28–29
  note-taking systems, 40–41
  paraphrasing, 34
  preparation/materials for, 30–32
  purpose of, 17
  in shorthand, 41–46, 44
  understanding vs. writing, 29,
    32–34
note taking while reading
  textbooks, 98–100

Peale, Norman Vincent, 187
phone distractions, 242–43
Pichert, James, 157–58
planning your work, 193–215
  block out daily time for learning, 198–200
  calendar use, 201–3
  developing a plan, 212–14
  and flexibility, 207
  for instructors, 214–15
  prospective memory needed for, 193–94
  reminders, 193–94
  setting and revisiting learning goals, 206–8, 214–15
  setting goals, factors in, 209–12
  and sleep, 195–98
  spacing your work out, 199–200
  to-do lists, 194, 204–6
  underestimating the time needed (the planning fallacy), 194, 200, 202
Plutarch, 16
Pomodoro technique, 255
praying before exams, 153
procrastination, 216–36
  accountability and shame as prods, 227–28
  breaking large tasks down into small ones, 221–22
  and impulse control, 217
  for instructors, 235–36
  and overestimating emotions, 226
  and pleasure/pain, 216–17
  prevalence of, 235
  progress tracking, 233–34
  reframing your choice as opportunity cost, 224–25
  relying on habit, 218–20
  relying on willpower, 220
  as self-handicapping, 230–31
  and self-image, 234
  starting, with permission to stop, 226–27
  support from friends, 228–30
  temptation turned into reward, 232–33
  to-do lists, 222–23
  and work streaks, 234
psychological distancing, 247

rereading, 140–41
reward and desire, 248–49

self-confidence, 259–72
  and career aspirations, 260
  in college vs. high school, 269–70
  comparing yourself to others vs. yourself, 266–68
  and family values re learning, 268–69
  feedback from instructors, 270–71
  "good learner," meaning of, 262–64
  importance of, 259–60
  for instructors, 270–72
  and self-image, 260–62, 266, 271
  and setbacks, 259–60
  via social support from good learners, 264–66
self-image, 234, 260–62, 266, 271

self-talk, 153–54, 249–50

shyness, 23–24

sleep, 195–98, 253

social contagion, 240

social media, enjoying vs. wanting, 248–50

speed reading and skimming, 101–2

study groups, 112, 124, 164, 228

study guides (flash cards), 111–14, 120–22, 142

support
    from friends, 228–30
    from good learners, 264–66
    visualizing someone supportive before exams, 154

tests. *See entries beginning with "exams"*

textbooks, 90–104
    contradictions in, 90–91, 103
    coordinating meaning across sentences in, 91–93
    deep reading of, 103–4
    headings and subheadings in, 99
    hierarchical format of, 92
    for instructors, 103–4

KWL (what you Know; what you Want to know; what you've Learned) approach, 97
    learning aids in, 101–2
    note taking while reading, 98–100
    posing/answering questions as you're reading, 97
    reading, allocating time for, 101–3
    reading/highlighting, 93–95
    reading strategies, 95–98, 103
    SOAR (Set goals; Organize; Ask questions; Record your progress) approach, 97
    speed reading and skimming, 101–2
    SQ3R (Survey; Question; Read; Recite; Review) approach, 95–96
    why they are difficult to read, 90

to-do lists, 194, 204–6, 222–23, 252, 255

tree diagrams, 78, 80

venting/chatting right before exams, 153

visualizing success or someone supportive before exams, 154

ABOUT THE AUTHOR

Dan Willingham received a PhD from Harvard University in cognitive psychology and spent over a decade researching how the brain changes as a consequence of learning. Now a professor of psychology at the University of Virginia, his bestselling first book, *Why Don't Students Like School?*, was hailed as "a triumph" by the *Washington Post* and "brilliant analysis" by the *Wall Street Journal* and was translated into many languages. His book *When Can You Trust the Experts?* was named recommended reading by *Nature* and *Scientific American* and made CHOICE's list of Outstanding Academic Titles for 2013. Willingham writes a column called "Ask the Cognitive Scientist" for the American Federation of Teachers' magazine, *American Educator*. He is a fellow of the American Psychological Association and of the Association for Psychological Science.